CONCEPTION, BIRTH AND CONTRACEPTION

CONCEPTION, BIRTH AND CONTRACEPTION
A VISUAL PRESENTATION

ROBERT J. DEMAREST

Department of Anatomy
College of Physicians and Surgeons
Columbia University

and

JOHN J. SCIARRA, M.D., Ph.D.
Department of Gynaecology
University of Minnesota

INTRODUCTION BY ANTHONY STORR

HODDER AND STOUGHTON

DEDICATION

This book is dedicated to our children, Robert, Steven and
Nancy Demarest and Vanessa, John and Leonard Sciarra

AUTHORS' NOTE

The growing demand for more and better sex education is evidence of a lack of satisfaction with things as they are.

Our contribution to this demand is a sequential text using, whenever possible, "life-size" illustrations. The power of an illustration to convey information directly is easily understood: it is the power to reinforce, transcend, or even alter the intent of the printed word. An illustration of any aspect of anatomy or physiology can impress a reader indelibly with what he reads. Herein lie both the value and the danger of an illustrated format. Poorly illustrated books have no place in basic instruction, particularly in basic education dealing with human reproduction.

In this book we have presented the fundamentals, the elementary knowledge needed to form the foundation for a mature understanding of human reproduction. We have tried to present this knowledge with a regard to aesthetics and human dignity. It has not been our intention to deal with the attitudes and moral qualities needed for this life. Undoubtedly, however, the information and manner of presentation will have an influence on these attitudes. An honest, straightforward survey such as this will, we believe, provide some of the necessary perspective.

Thanks are due to Ortho Pharmaceutical Corporation for their cooperation, and to Frank Taylor, Editor-in-Chief of McGraw-Hill's Trade Division, who from the very beginning has enthusiastically encouraged and guided us through the many months of preparation. We would also like to express our gratitude to Dr. George Langmyhr of Planned Parenthood Federation of America, who reviewed the material on contraception.

Robert J. Demarest
John J. Sciarra, M.D., Ph.D.

CONTENTS

FOREWORD

When I was a medical student, some twenty-five years ago, I was sent to a maternity hospital to learn midwifery. Most of the mothers whose babies I helped to deliver have entirely faded from my memory; but one girl created a vivid impression. She was a redhead, and it was her first baby. Her labour was progressing perfectly normally, but it was obvious that she was much more apprehensive than was usual. Indeed, her eyes were dilated with fear. I tried to reassure her, but it was useless. Finally, she revealed what was alarming her. "When does it happen?" she asked. "When do I split down the middle?" Although actually in labour, she had no idea of the process of birth, or even of her own anatomy. She believed that, at some point, her abdomen would suddenly be rent in two, and that the baby would emerge much as actually happens with Caesarian section; and, rather naturally, she imagined that this splitting would be excruciatingly painful. In those days, I had neither the knowledge nor the time to explore the origin of her fantasy, and discover why she had so thoroughly repressed all awareness of her genital anatomy. I know now that such girls are generally frigid, and that their ignorance is motivated by guilt and fear. Even if the very best information and instruction is available, there may still be a few girls who are unable to face the reality of their own bodies; but I venture to think that the far healthier attitude which nowadays prevails has reduced their number to a minimum. This attitude is reflected in the many books available on sex and allied topics.

"*Conception, Birth, Contraception*" is a notable addition to the literature on these important subjects. It is exactly the right kind of book to give an adolescent, or to leave on the family bookshelf to be consulted when needed. There is no moralizing, no advice, no speculation. What is presented is a clear account of the facts of reproduction so far as we understand them, and a series of illustrations showing what organs are concerned in the process and how they function. If my red-headed young patient had been given this book she would have been spared a great deal of unnecessary anxiety. She would also have found it fascinating to see pictures of exactly what was going on inside her own womb during the long nine months of pregnancy.

The joint authors are American: a gynaecologist and an

anatomist. The fact that one is an anatomist no doubt accounts for the beautiful accuracy of the illustrations, which are life-size whenever this is possible. I know no other book on reproduction which is so admirably illustrated. The text has been slightly modified in a few places where American usage might cause confusion to English readers. Unlike many American texts it is clear, concise and readable. In my view, the book will be found valuable by teachers, and should be obligatory reading for teenagers before leaving school. There are, of course, some long words. An accurate scientific account, which this book is, cannot avoid using a few technical terms. But this should not deter teachers from using it. They will find that their pupils are fascinated, and that those who are unlikely to remember or spell such words as "endometrium" will nevertheless gain a clear impression of what reproduction is all about from the pictures. Nor should it be imagined that only the young will benefit. One friend to whom I showed it was enthusiastic. "It's just what I need," she exclaimed. "There's still a lot I don't know about the whole process." And she happens to be a highly intelligent architect who has had four children.

Anthony Storr

CONCEPTION, BIRTH AND CONTRACEPTION

CONCEPTION

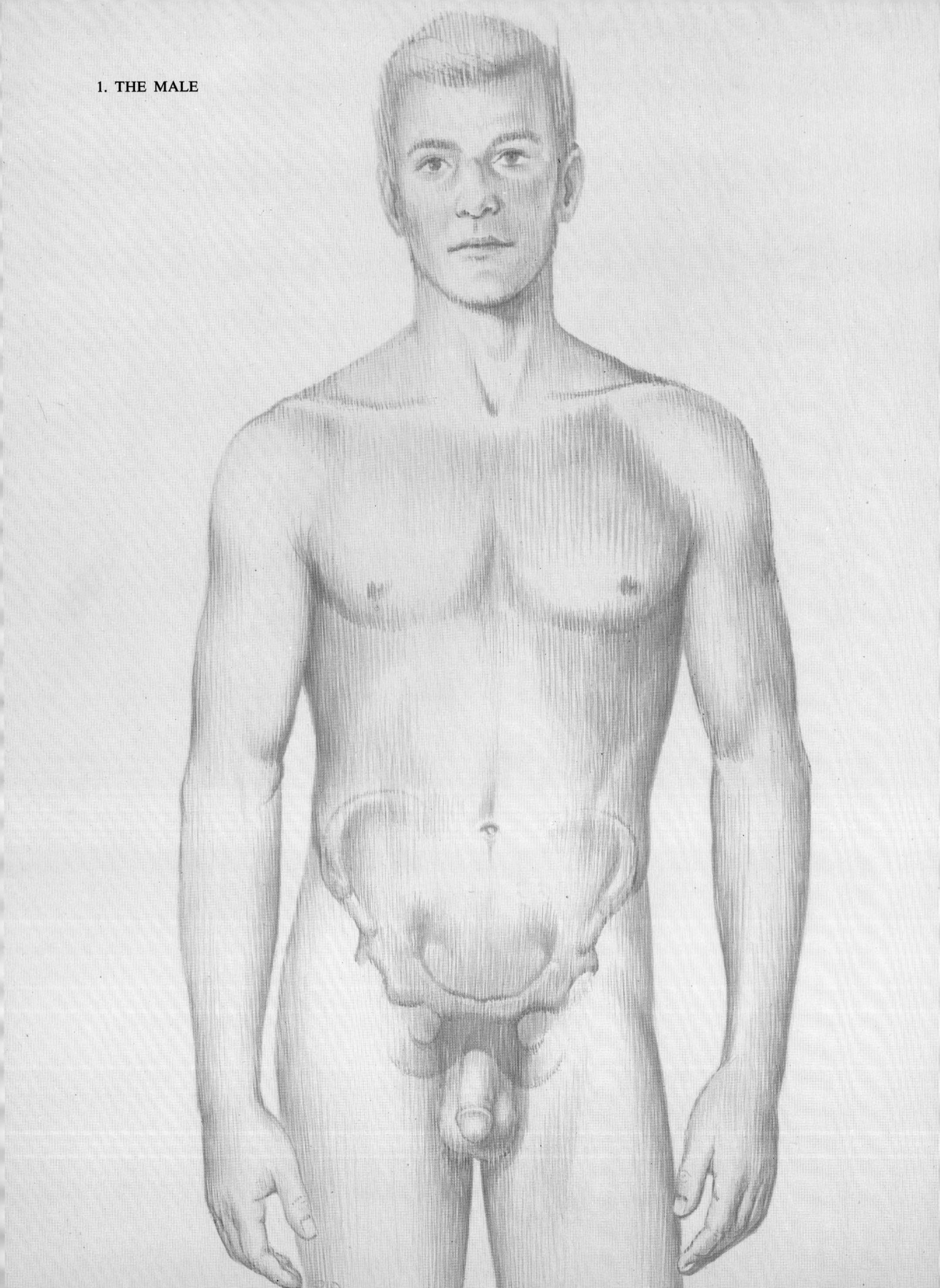

1. THE MALE

THE MALE

THE ROLE AND FUNCTION OF THE MALE IN REPRODUCTION

In the reproduction of human life, the female plays a prolonged and fascinating role. She provides the egg which, if fertilized, gives rise to a new life. Her body shelters the growing foetus and supplies the food and oxygen that it needs to develop into a full-term baby. From fertilization to delivery, mother and child are as one for approximately 266 days. At the end of this period the mother delivers the infant into the world and nurses it through its perilous early life.

The role of the male in reproduction may seem less important than that of the female. He need only generate the spermatozoa and deliver them to the cervix, deep within the vagina a process which requires only a few brief minutes. Yet, without the sperm there can be no fertilization and without fertilization there can be no new life.

The tiny sperm that fertilizes the egg carries with it chromosomes which, when combined with those in the egg, determine all the inherited characteristics of the baby. The sperm also contains the single chromosome that determines the sex of the baby. Unless the sperm reaches the egg at just the right time, fertilization will not take place and the egg will deteriorate and be expelled.

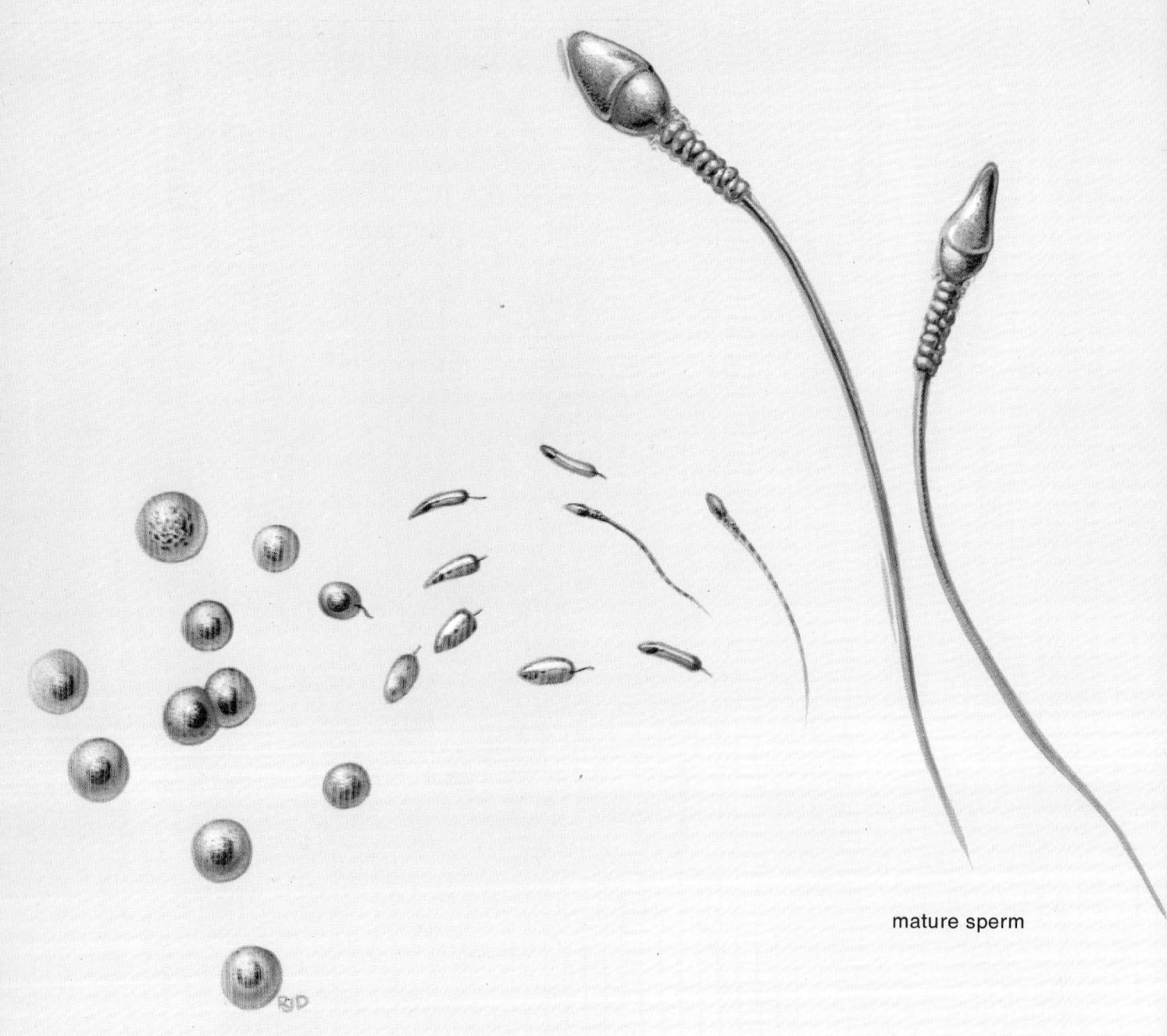

mature sperm

spermatocytes

Because it is such a tiny living cell the sperm cannot be seen without a microscope, and it would take many thousands of spermatozoa just to cover the dot at the end of this sentence. What the sperm lacks in size, it makes up for in numbers and ability to move. On ejaculation, which occurs during sexual intercourse, as many as five hundred million spermatozoa may be expelled. These mature, motile sperm can travel at the speed of $\frac{1}{8}$ inch per minute—a very rapid rate for an organism so small.

Illustration 2 shows the process by which a spermatocyte develops into a mature sperm with a head and tail (here magnified several thousand times). This maturation process, called spermatogenesis, takes between sixty and seventy days.

At any given time, millions of sperm are being produced within the male reproductive system. This is quite different from the female reproductive system in which usually only one egg matures each month and is available for fertilization during a very brief period.

THE MALE
REPRODUCTIVE SYSTEM

As in all of nature, the structure of a body system and its function are always in perfect harmony. The male reproductive system, which is responsible for producing the sperm, is no exception. It is, for the most part, located at the outlet of the bony pelvis in an area called the perineum. Its location outside the main confines of the body is understandable when one considers that optimum sperm production requires a temperature lower than that found inside the body. The male pelvis as shown in Illustration 1 is heavier than that of the female, more funnel-shaped and less roomy. It seems to be designed for weight bearing, whereas the female pelvis, being larger and more basin-shaped, is designed for childbearing.

The testes, the paired genital glands of the male, are oval, somewhat flattened structures, each about $1\frac{1}{2}$ inches in length. They are contained in the scrotum, or scrotal sacs—the skin pouches, which, together with the penis, make up the visible external reproductive organs. In addition to producing the spermatozoa, the testes secrete the male sex hormone testosterone. The left testis, or testicle, is usually lower than the right, and the left scrotal sac, accordingly, lower and slightly larger. The testes in the male, like the ovaries in the female, are formed within the abdomen during the development of the embryo. Prior to the male child's birth, the testes descend into the scrotal sacs. Occasionally, a testicle does not descend, but remains in the abdominal cavity or in the canal that traverses the abdomen. Such a condition should be corrected before the male reaches maturity.

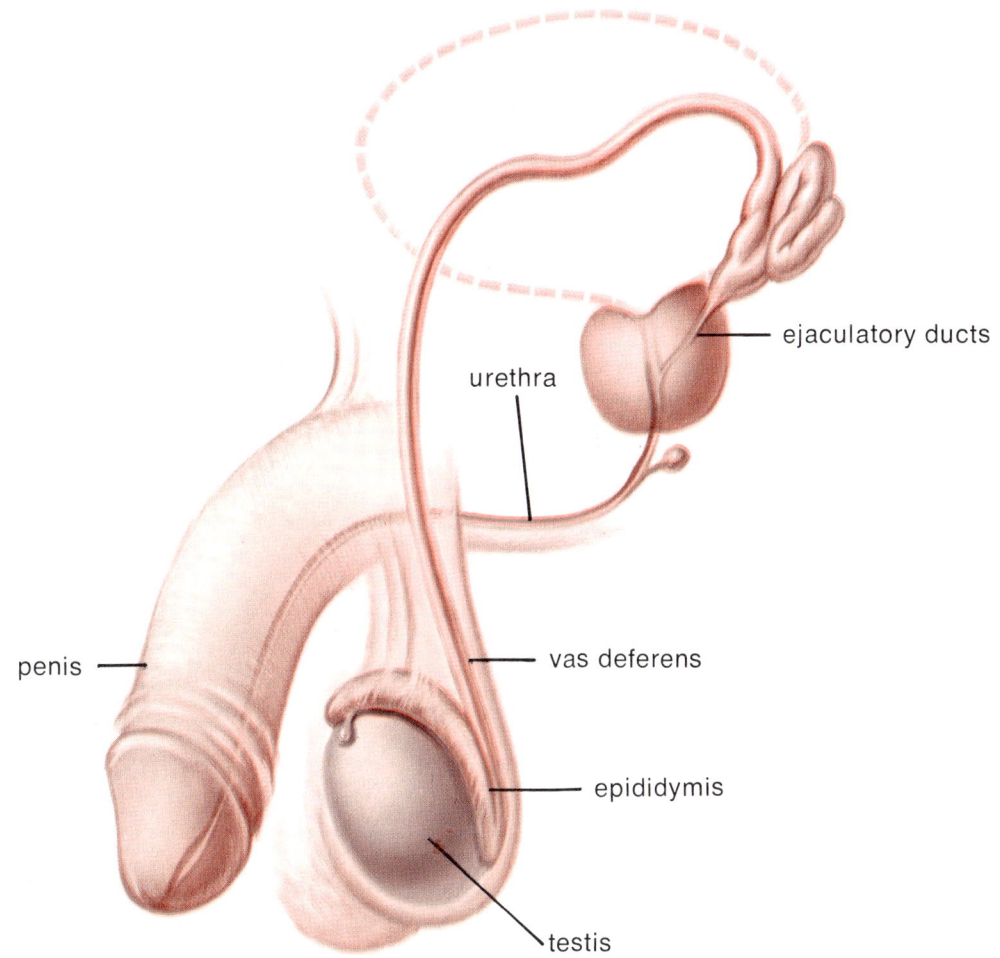

ejaculatory ducts

urethra

penis

vas deferens

epididymis

testis

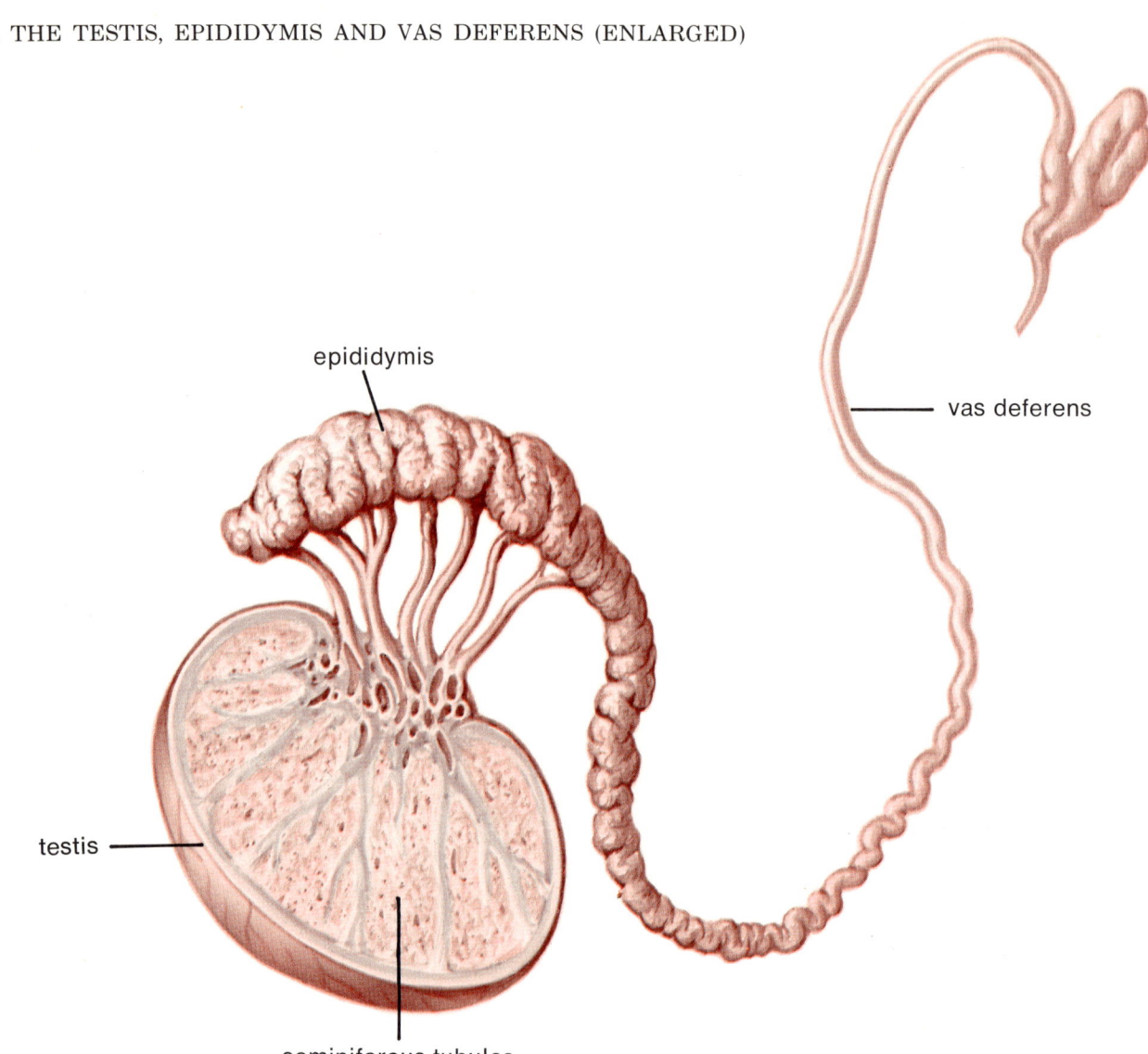

epididymis

vas deferens

testis

seminiferous tubules

Adjacent to each testis is an elongated structure called the epididymis. The sperm reach the epididymis through a series of ducts. In the epididymis they undergo final maturation, which takes only a few hours.

A *vas deferens* is attached to each epididymis. The *vas deferens,* together with the blood vessels and other tissues surrounding it, is called the spermatic cord. The route the sperm must take is shown in Illustration 4. Sperm travel from the testis, into the epididymis, and from there through the *vas deferens* of the spermatic cord. The *vas deferens* from each testis winds upward over the pubic portion of the bony pelvis and into the pelvic cavity. Here the two cords come together and form side by side midline tubular structures about 1 inch in length. These structures, called the ejaculatory ducts, connect directly with the urethra close to the neck of the urinary bladder. The urethra, or urinary duct, is an S-shaped tube about 8 inches long, extending underneath and through the penis. It further transports the sperm out of the body of the male by way of the penis. Thus it can be seen that in the male, the reproductive and urinary systems are intimately related.

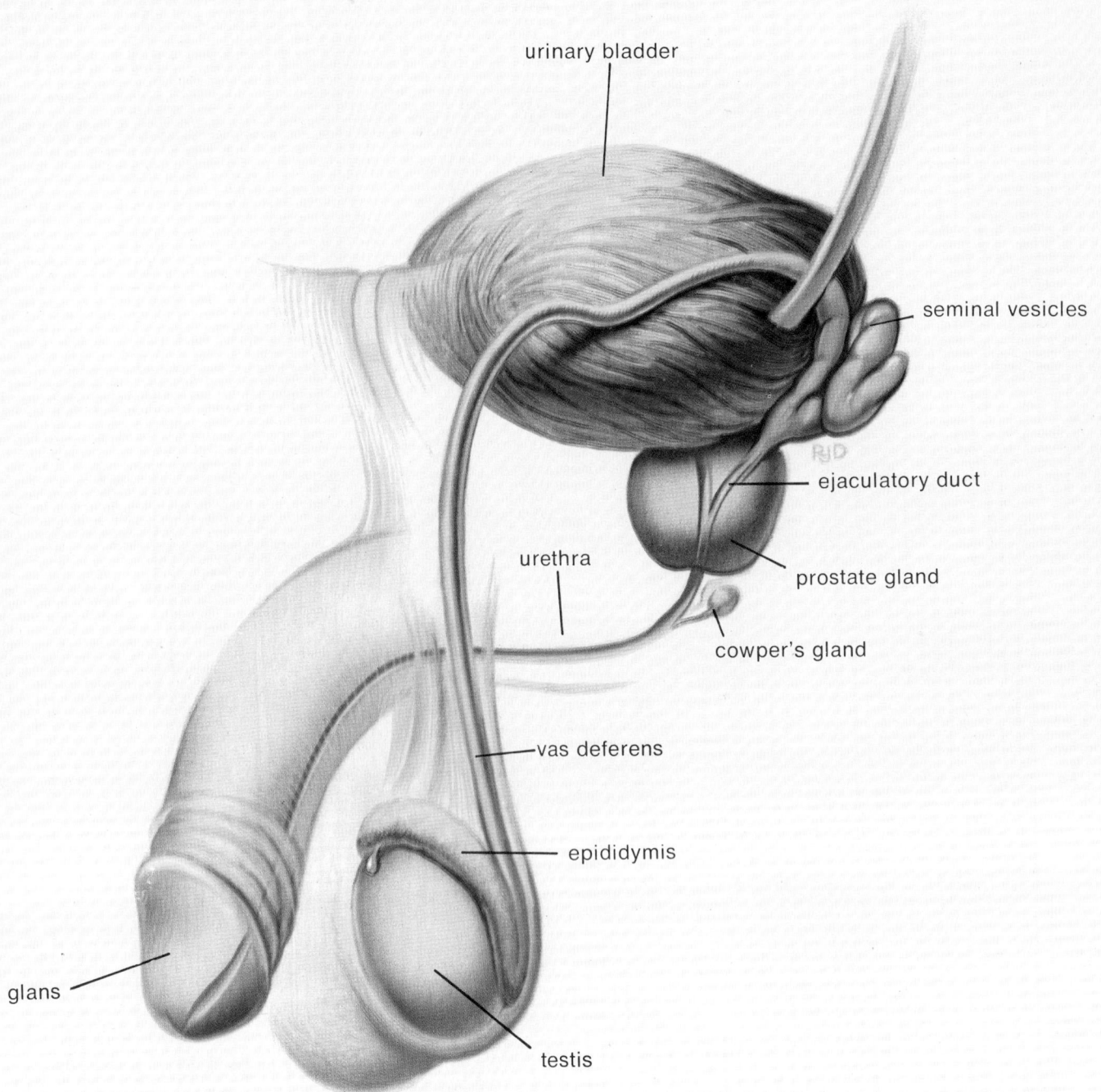

urinary bladder

seminal vesicles

ejaculatory duct

prostate gland

urethra

cowper's gland

vas deferens

epididymis

glans

testis

As the sperm move through the male reproductive system, the secretions from three sets of glands produce a thick, milky substance which helps transport them. The paired seminal vesicles were, at one time, thought to store sperm, but it now appears that their chief role is to produce a secretion which is added to the sperm as they enter the ejaculatory ducts. The tiny Cowper's glands (or bulbourethral glands), located on either side of the urethra, secrete a mucous substance into the urethra which also helps with sperm transport. The prostate gland, located at the base of the urinary bladder where it joins the urethra, is by far the largest of the three glands discussed here. The prostate is about $1\frac{1}{2}$ inches in diameter, composed of glandular tissue and, like the seminal vesicles, produces a milky fluid which accompanies the emission of sperm. In older men, the prostate sometimes enlarges and obstructs the flow of urine through the urethra.

At the time of orgasm during sexual intercourse, about 1 teaspoonful of grey-white fluid called semen is ejaculated. The semen, or seminal fluid, contains the sperm and the secretions from the seminal vesicles, the prostate, and the Cowper's glands. If the male has not had an orgasm in several days, as many as five hundred million sperm may be present in the semen. If he has had repeated recent ejaculations, however, the volume of seminal fluid and the number of sperm expelled will be reduced.

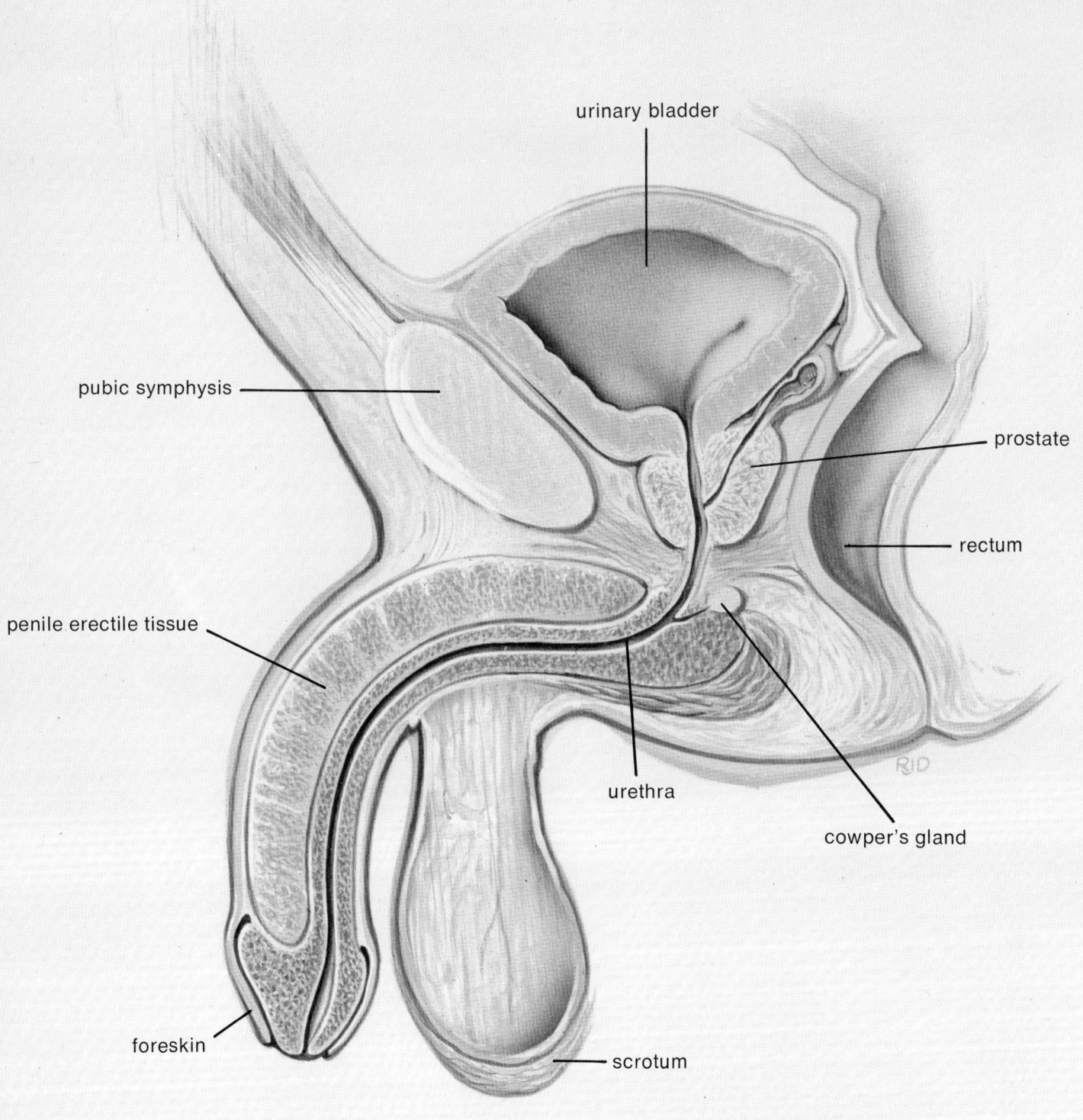

urinary bladder

pubic symphysis

prostate

rectum

penile erectile tissue

urethra

cowper's gland

foreskin

scrotum

To deliver the sperm to the cervix of the female uterus, the male's penis must change from a relatively short pendulous organ to a larger, erect structure.

At birth, the glans penis is covered by a loose fold of skin—the prepuce or foreskin. Soon after birth, this foreskin is sometimes removed in an operation called circumcision. The removal of the foreskin prevents possible constriction of the penis and resulting restriction of circulation which could be injurious. It also prevents the accumulation of any secretion under the foreskin which might cause inflammation or infection. Illustration 6, which presents a cross section of the male pelvic organs shows the foreskin of the penis still present. All other illustrations in this book show it removed.

In its erect condition, the penis can penetrate the vagina and expel the sperm deep within the female reproductive system. The size of the flaccid, nonerect penis is not specifically related to the general physical size of the male, nor is it directly related to the size of the penis during erection. In some males, the organ may increase only slightly; in others, it may almost double in overall length. The penis of an adult male, which measures 4 inches in the flaccid, pendulous state, may increase in length by 3 inches in the erect condition. There is no bone in the human penis, although some animals do have a penile bone.

Erection of the penis is controlled by nerve impulses from the spinal cord. These nerve impulses dilate the arterial blood vessels in the penis, allowing the erectile tissue in the shaft of the organ to fill with blood. While the arteries are filling with blood, the veins of the penis are being compressed, building up pressure within the erectile tissue and causing the organ to become firm and erect. The clitoris of the female is also composed of erectile tissue and undergoes similar changes during sexual stimulation, although the changes are not so obvious or so pronounced as in the penis. The clitoris is discussed in greater detail in the section on the female reproductive system.

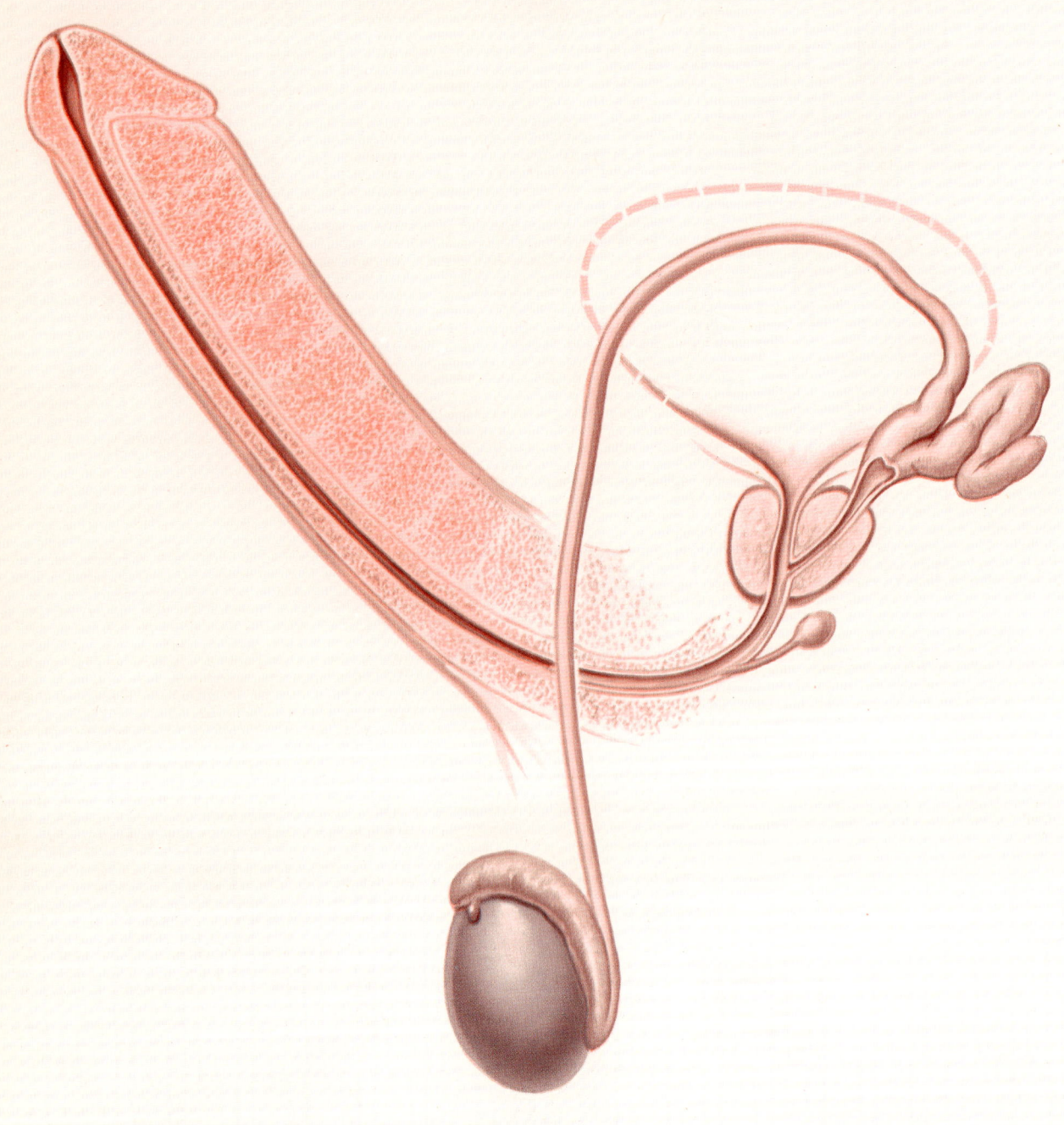

While the penis is in the erect condition, the second activity necessary for delivery of the sperm occurs—the discharge of semen, containing the sperm, from the penile urethra. The discharge occurs at the time of male orgasm. Actually, two processes take place under the control of the male nervous system. One is emission and the other ejaculation. Before ejaculation can occur, the sperm must move from the epididymis where they are stored, along the *vas deferens*, and through the ejaculatory ducts into the urethra. The process by which the sperm move through this part of the male reproductive system is emission. At this point, the sperm and the secretions from the male reproductive glands are violently propelled along the urethra and expelled at the terminal end of the penis. In the true physiological sense, this action is referred to as ejaculation. The contraction of the muscles along the base of the penis and the urethra provide the force to discharge the semen at the time of ejaculation.

At the completion of the sexual act, the veins and arteries of the penis return to their normal state, the excess blood drains out of the erectile tissue, and the organ assumes its flaccid nonerect size.

Illustration 7 shows the path of the sperm from the testes through the male reproductive tract. The penis is illustrated in the erect position, showing the increased distance the sperm must travel through the penile urethra at the time of ejaculation.

After ejaculation, the male has completed his role in the phenomenon of reproduction. He has generated the sperm and delivered them to the cervix of the female. Somewhere between one hundred million and five hundred million sperm are now ready to move through the female reproductive system to fertilize the egg. The attrition rate of the sperm is extremely high, but if the environment within the female reproductive system is right, at least one will reach the egg, fertilize it, and conception will take place. A new life will begin.

PUBERTY AND THE EFFECT OF HORMONES ON THE MALE REPRODUCTIVE SYSTEM

The reproductive system of the male does not become functional until about the age of twelve years. At this time, the pituitary gland at the base of the brain increases the amount of gonadotrophic hormones that it secretes into the bloodstream. These are carried to the testes and stimulate them to begin the production of testosterone, the male sex hormone.

The presence of testosterone causes changes in the male's body. His genital organs enlarge and hair begins to grow on his face, under his arms, and in the pelvic region. Further, the testosterone and the gonadotrophic hormones act on the tubules of the testes, causing them to begin the production of sperm. Once sperm production begins in the male, it continues throughout life.

The time when these changes begin to take place is called puberty.

THE FEMALE

THE ROLE AND FUNCTION OF THE FEMALE IN REPRODUCTION

People generally think of human reproduction only in terms of the female's role. This is not too surprising since it is the female who carries the baby during the nine months of its prenatal life. Her body houses the foetus, maintaining the ideal environment for its growth, protecting it from injury, and providing it with nutrients. For many months before delivery, changes in her physical appearance announce the impending birth. And with the birth of the baby, her role as nurse and mother assumes primary importance in both her life and that of her offspring.

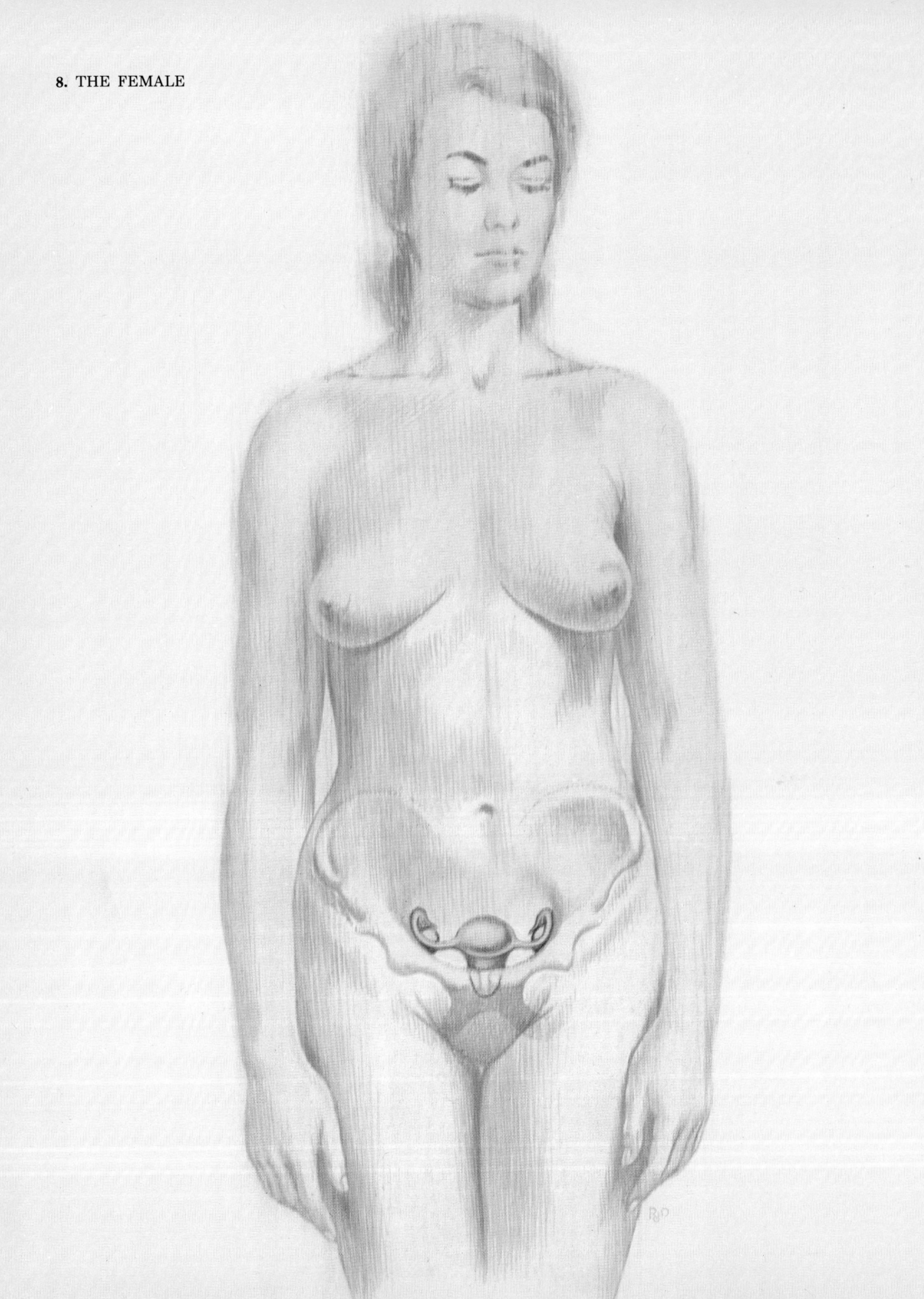

8. THE FEMALE

THE FEMALE REPRODUCTIVE SYSTEM

As with the male, the female body is constructed to serve its role and function in reproduction.

The reproductive system of the female is composed of both internal and external organs. The major reproductive organs of the female are located within the body. The external genital organs and the breasts are of only minor importance in reproduction. During pregnancy, the breasts enlarge and prepare to secrete the milk that will nourish the newborn infant. The genitals serve as the terminus of the female reproductive tract, providing the opening through which the sperm must pass if fertilization is to occur and, later, through which the child is delivered. The anatomy of the female reproductive system will be shown in several subsequent illustrations.

Collectively, the external female genitals are referred to as the vulva. The clitoris is a small erectile organ covered with a fold of tissue, the prepuce, which is much like the foreskin of the penis. Structurally, the clitoris is similar to the penis, but much smaller. It serves no specific physiologic function in reproduction but does contain tactile receptors which when stimulated help arouse the female sexually during intercourse.

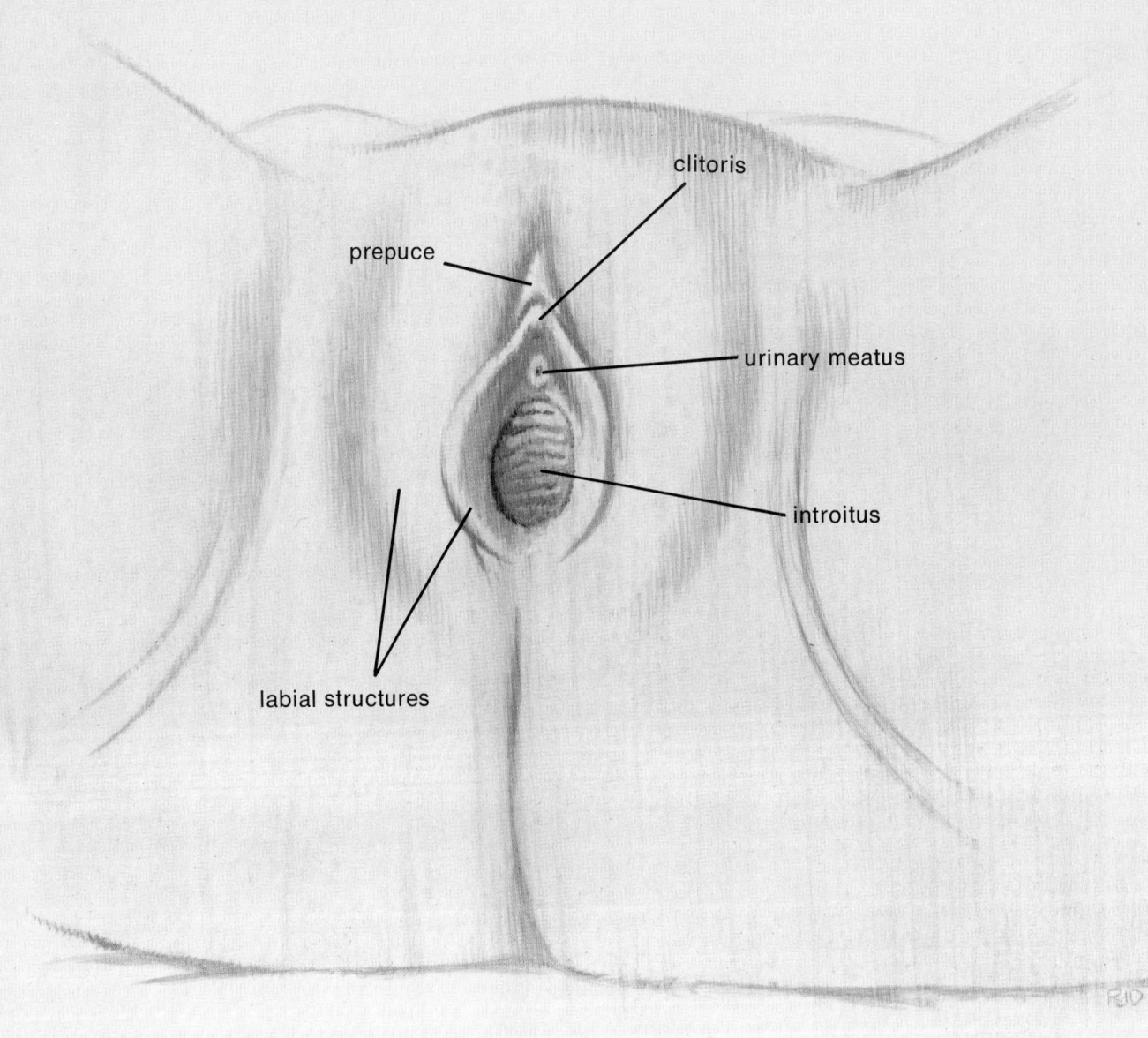

clitoris

prepuce

urinary meatus

introitus

labial structures

Just beneath the clitoris are the labia minora and majora which surround the vaginal opening, or introitus. Within the vaginal opening is the vaginal canal. In the illustration, the wall of the vagina appears wrinkled because the vagina is in a relaxed state. When it is expanded, the convolutions disappear and the vaginal wall appears smooth.

Just above the vaginal opening is a smaller opening called the urinary meatus, which is the external opening of the female urinary tract. In the female, the urinary and reproductive systems are separated. In the male, the urethra is used for the transport of both urine and sperm, although never simultaneously.

In the young female, the vaginal opening is partially covered by a thin ring of tissue called the hymen. The hymen is stretched during the teen years if tampons are inserted into the vagina to collect the menstrual discharge. Sexual intercourse also stretches the hymen, and in the mature female there usually remains only a small amount of hymeneal tissue. Occasionally, however, the hymen remains rigid and intact, blocking the entrance to the vagina. When this happens, a doctor may be called upon to stretch the hymen or cut part of it away.

10. INTACT HYMEN AS IT APPEARS IN THE YOUNG FEMALE

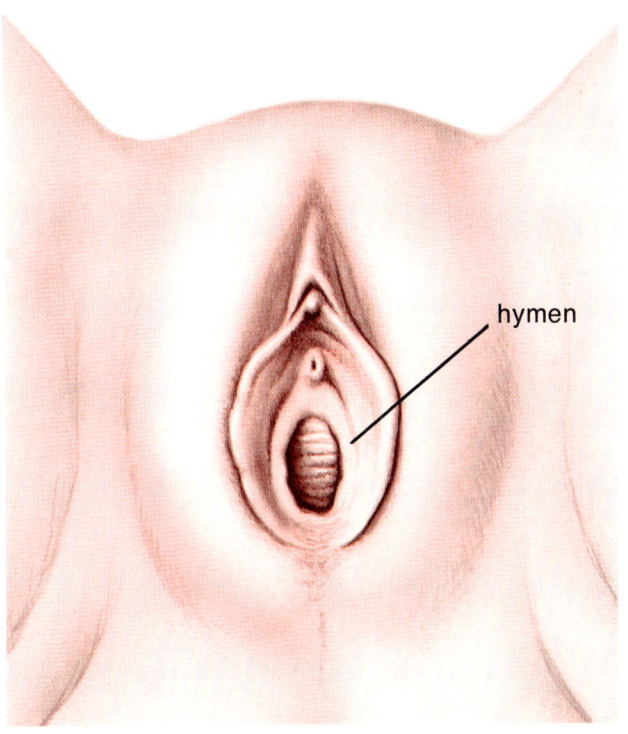

hymen

The internal reproductive organs of the female play the major role in reproduction. These organs are located in the pelvic portion of the abdominal cavity.

The bony pelvis of the female is quite different from that of the male. The male's is funnel-shaped, whereas the female's is basin-shaped and has a much larger outlet. The opening at the base of the female pelvis permits passage of the baby at birth.

Within the pelvic region are located the vagina (or vaginal canal), the uterus, the two fallopian tubes (or oviducts), and the two ovaries (Illustration 11 shows the position of these organs). The ovaries, in addition to producing eggs, secrete and release female hormones which act to control the reproductive cycle and also produce some of the female's secondary sexual characteristics—to be discussed in a subsequent chapter.

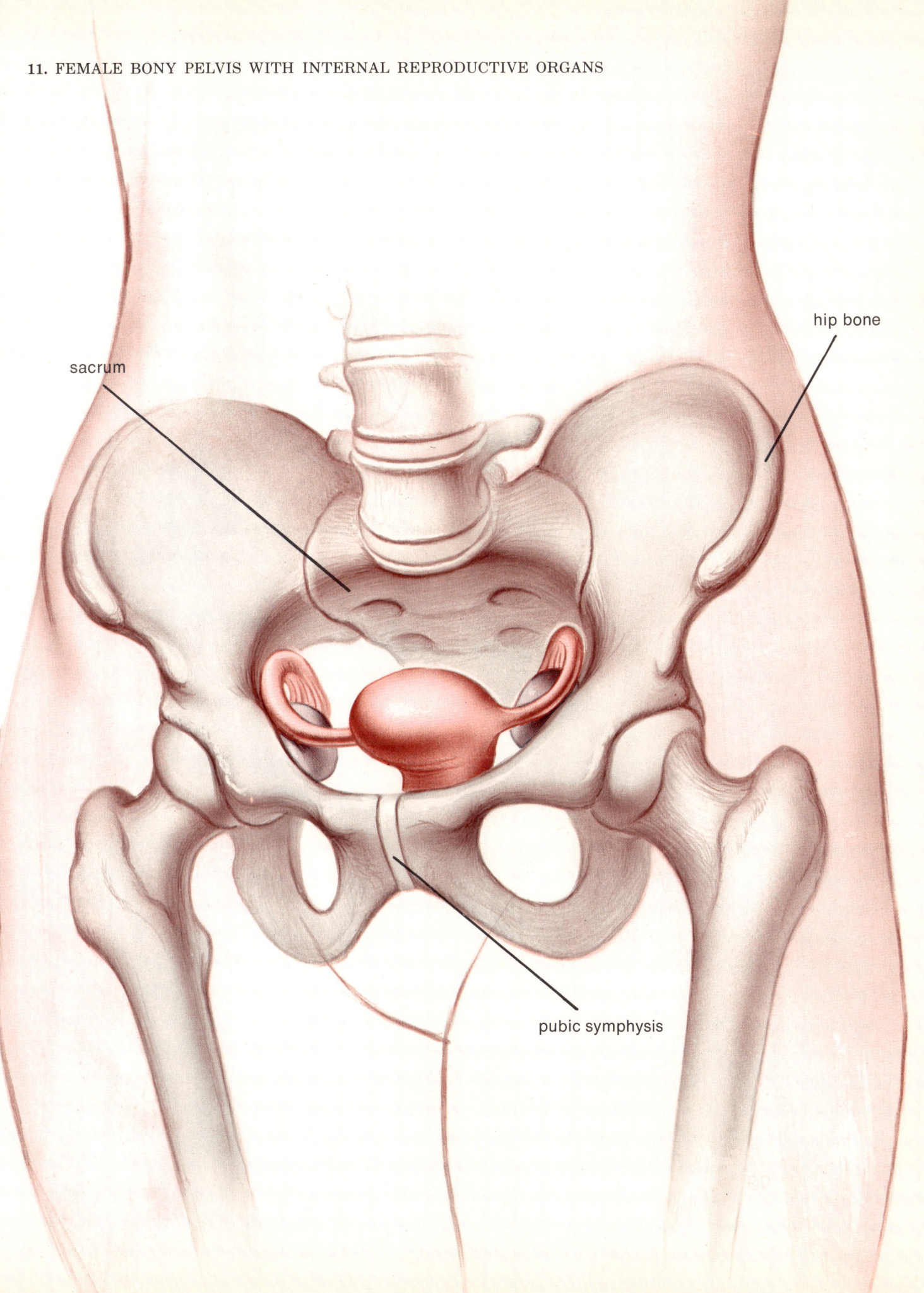

sacrum

hip bone

pubic symphysis

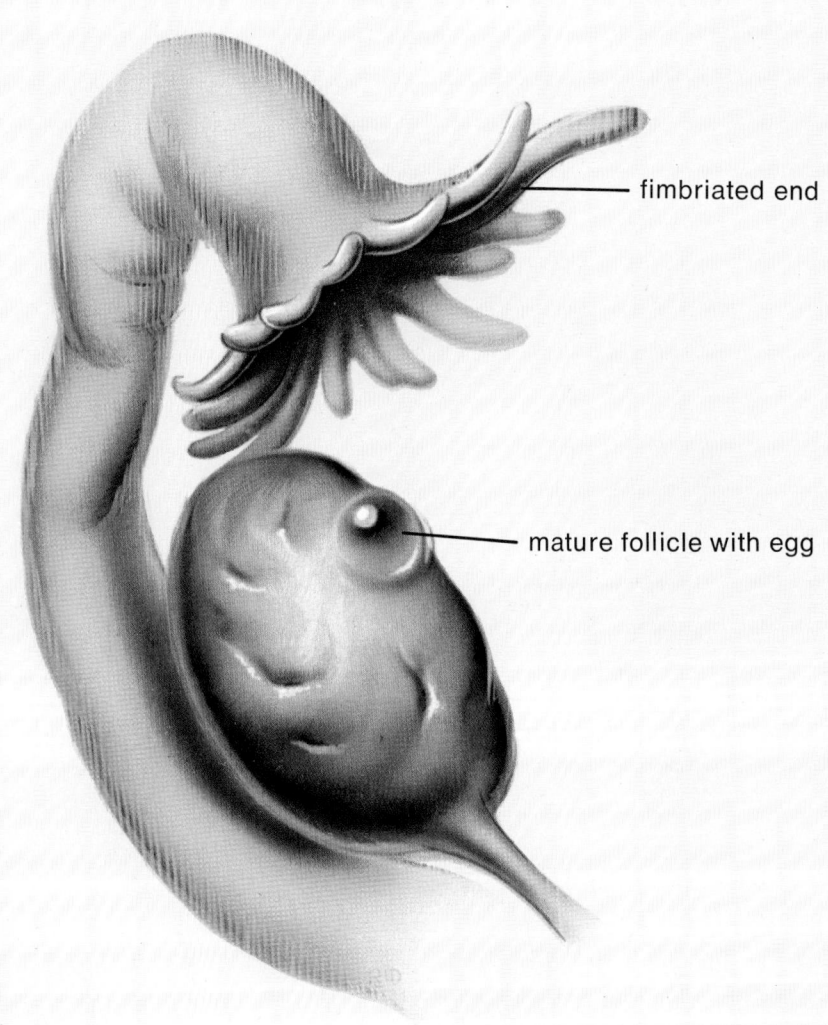

fimbriated end

mature follicle with egg

The female's ovaries are each about 1½ inches in length. They are flattened, whitish structures with a pitted surface, and are located adjacent to the fallopian tubes and uterus as shown in Illustrations 12 and 13. At the birth of the female her ovaries already contain all the eggs that will ever be released during her reproductive years. It is estimated that there are over one hundred thousand eggs present at birth, but usually only one egg is released each month. In all, only about four hundred eggs are successfully ovulated during the life of a normal woman. The remainder degenerate, most at the termination of the reproductive years—the time in the female's life called the menopause.

Ovulation, the process by which an egg is released, is a fascinating one. Throughout each ovary are distributed tiny groups of cells called follicles. Within each follicle is an immature egg or oöcyte. As a single follicle matures, it grows in size and emerges through the ovary's surface. Once every twenty-eight days a mature follicle ruptures and releases an egg.

Whereas in the mature male sperm are always available for fertilization, the fertile period of the female is cyclic, occurring for only a few hours once every twenty-eight days. During ovulation—the period of the release of the egg—the end of the fallopian tube closer to the ovary surrounds most of the ovary and its follicle. Once released from the follicle, the egg enters the adjacent fallopian tube, which will then move it to the uterus. The egg is not able to move by itself, but is transported by the action of the cells lining the inside of the tube and by contractions of muscle fibres within the wall of the tube. The distance from the ovary to the uterus is about 4 inches, and the journey of the egg requires several days. As stated before, it is only during this journey that the egg can be fertilized by the sperm which ascend the tube and meet it.

13. OVARY WITH DEVELOPING FOLLICLES (ENLARGED)

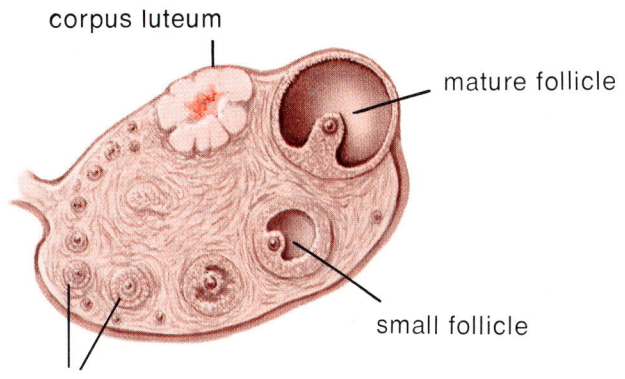

corpus luteum

mature follicle

small follicle

primary follicles

The mature egg is barely visible to the naked eye, and is only about the size of the full stop at the end of this sentence. However small it may be, the egg is about two thousand times as large as the sperm that must fertilize it. One reason for its larger size is that it carries the food the growing embryo will use during the first few days of its life.

Sperm can survive within the female reproductive system for several days. If intercourse has previously occurred, there may be living sperm already present in the fallopian tube when the egg enters, or the sperm may enter the uterus end at approximately the time the egg enters from the ovary.

The egg slowly passed down the tube while the mobile spermatozoa move up the tube to meet it. Fertilization usually takes place in the midportion of the tube. Usually, only one of the many spermatozoa present fertilizes the egg.

Illustration 14 shows a magnified cross section of the fallopian tube, the many convolutions of the inner lining, and the egg in the central channel of the tube.

14. CROSS SECTION OF FALLOPIAN TUBE (ENLARGED)

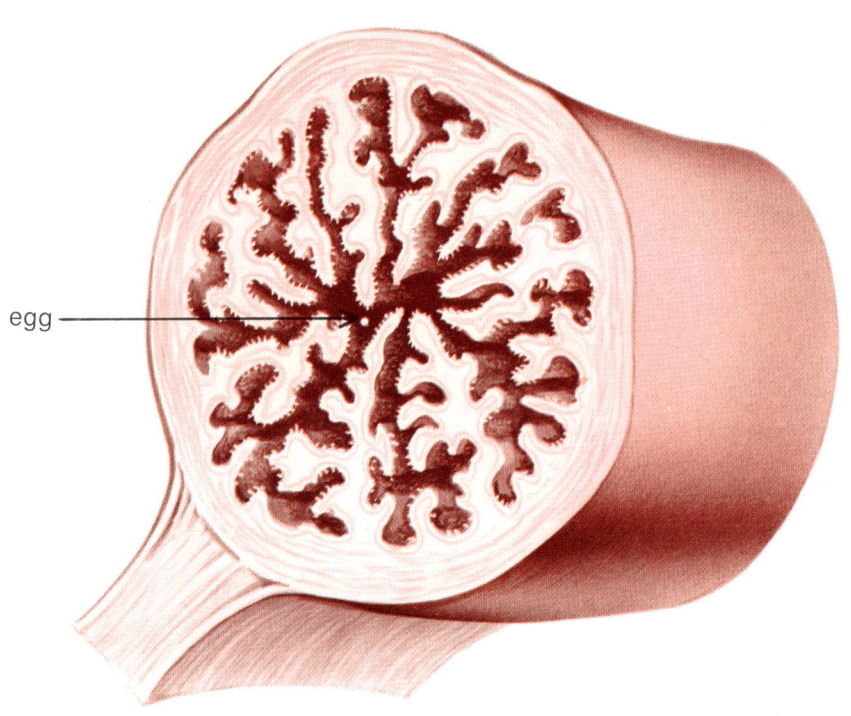

egg

15. CROSS SECTION OF UTERUS, TUBE AND OVARY (ACTUAL SIZE)

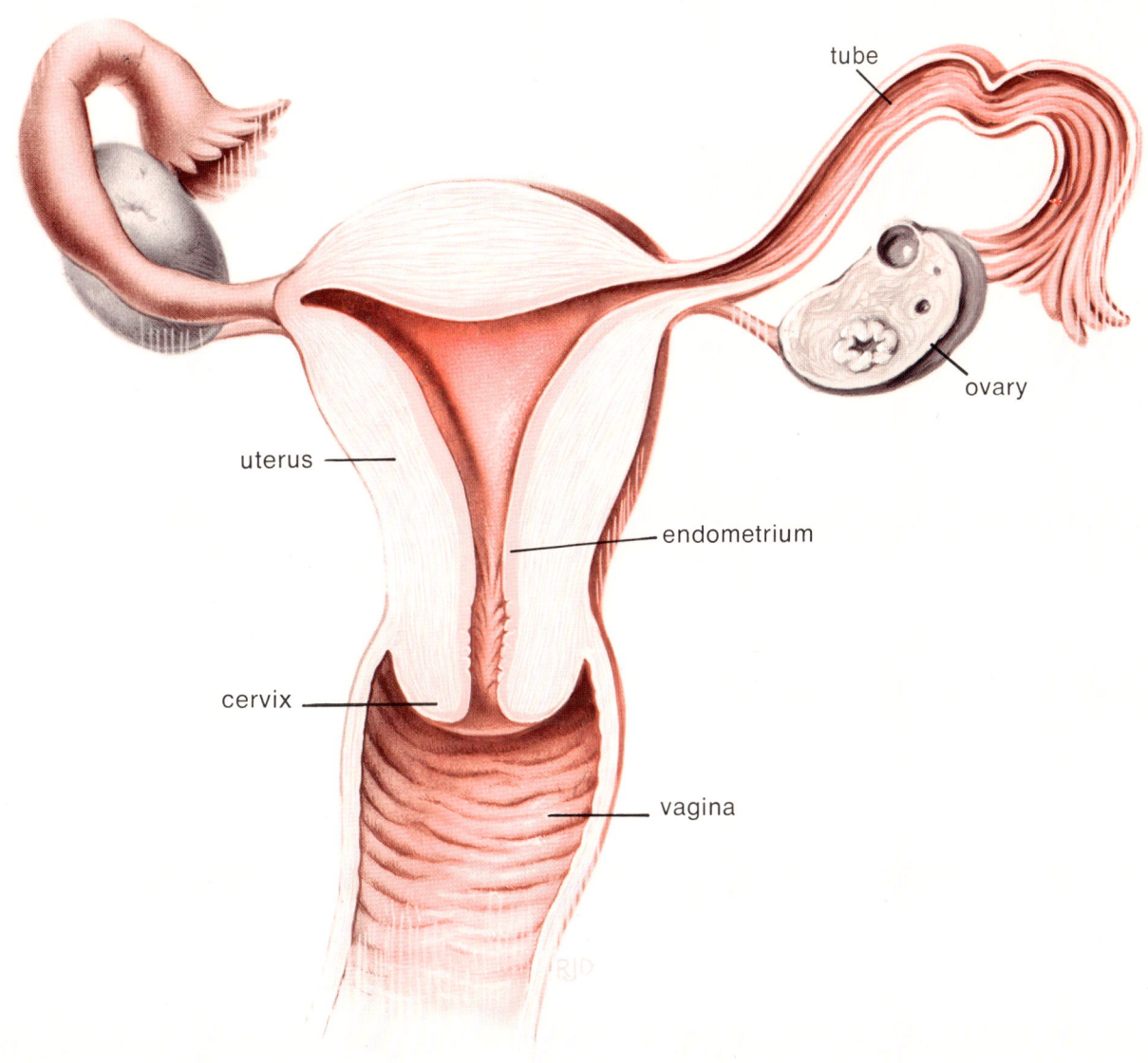

tube

ovary

uterus

endometrium

cervix

vagina

The egg, having now been fertilized in the fallopian tube, continues to move slowly down the tube toward the uterus. The uterus is a hollow, muscular organ which, in the nonpregnant state, measures about 3 inches in length. Its location within the body cavity is shown in Illustrations 8 and 11. The primary function of the uterus is to serve as the place where the fertilized egg will grow and develop.

The upper and lower part of the uterus, called the corpus or body, is normally flexed forward over the dome of the urinary bladder. The lower portion, called the cervic or the mouth, protrudes slightly into the vaginal canal. It is along the canal within the cervix that the sperm must pass on their way to the site of fertilization in the midportion of the fallopian tube.

The canal within the cervix connects at one end with the vaginal canal and at the other with the fallopian tubes, which lead to the ovaries. The inner lining of the uterus, called the endometrium, plays an important role in the reproductive (or menstrual) cycle of the female. During the course of each menstrual cycle this lining undergoes rapid growth. At the cycle's end, if fertilization does not take place, menstruation occurs, during which the accumulated cells are sloughed off, and the lining reverts to its pregrowth state. If fertilization does take place, the fully developed lining of the uterus is available for implantation of the fertilized egg.

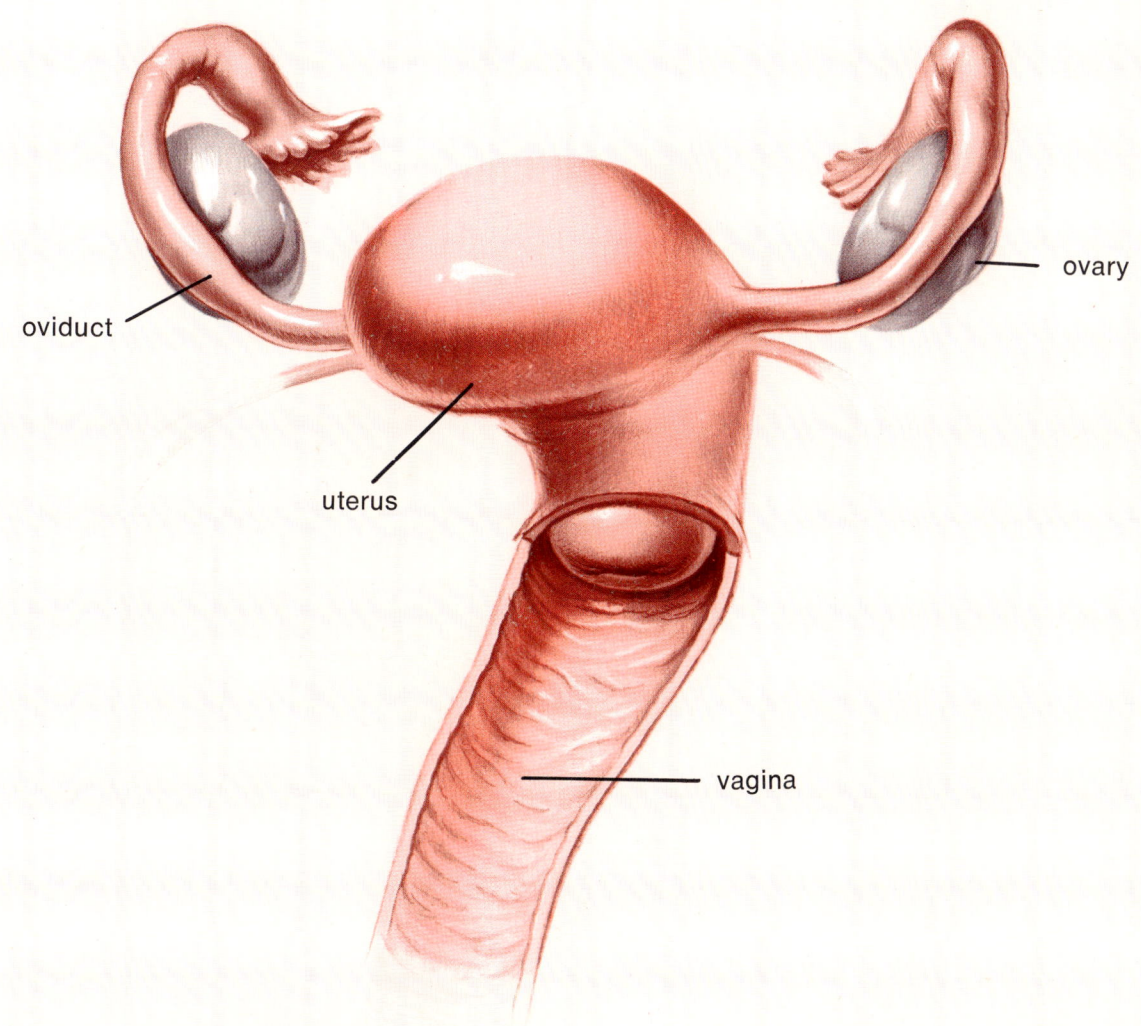

oviduct

ovary

uterus

vagina

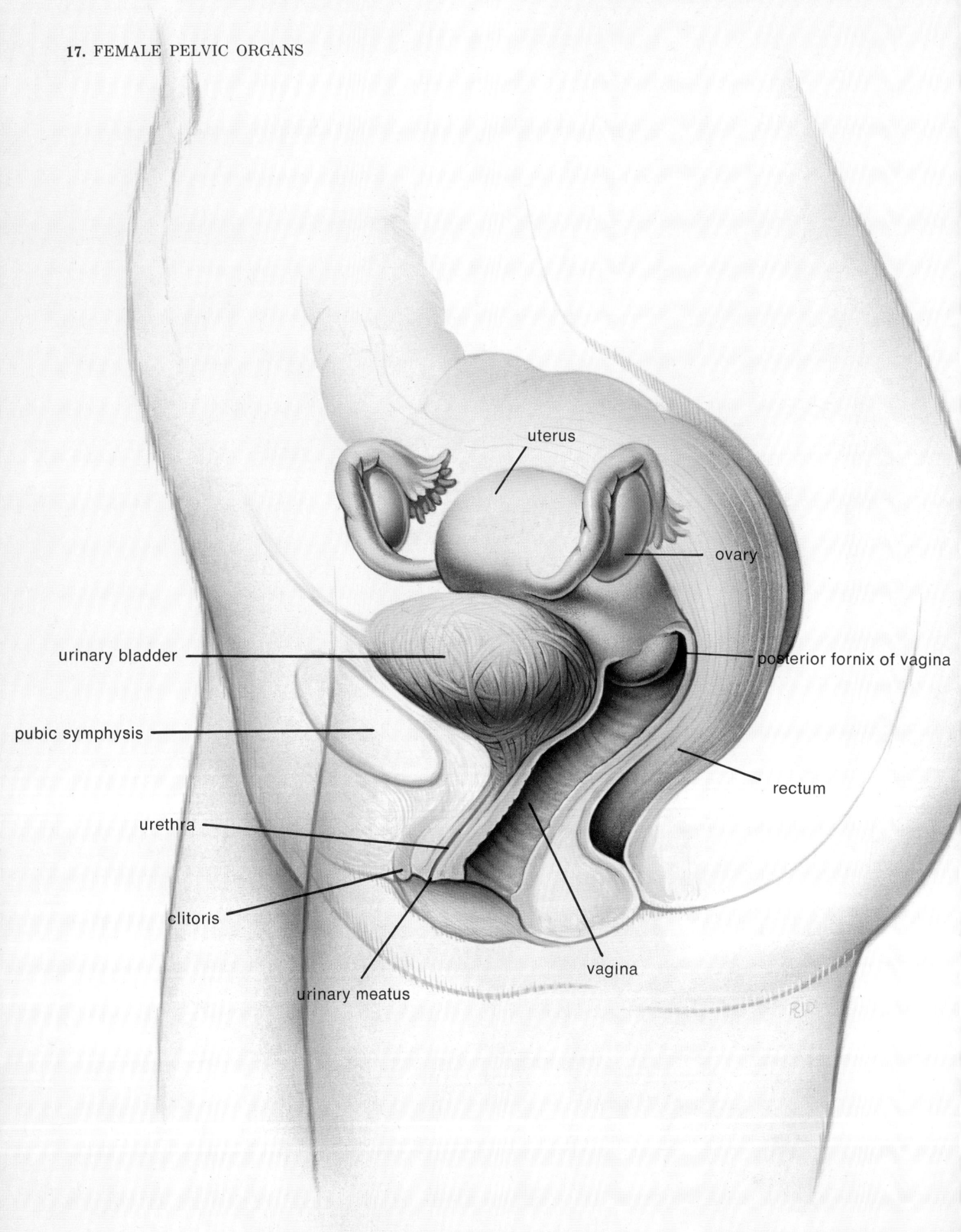

uterus

ovary

urinary bladder

posterior fornix of vagina

pubic symphysis

rectum

urethra

clitoris

vagina

urinary meatus

The first requirement of all for fertilization is that during the act of sexual intercourse (or in rare instances, by artificial means) semen from the male is deposited within the vaginal canal of the female. The vagina is a tubular organ about $3\frac{1}{2}$ inches in length, extending from the vulva to the cervix of the uterus. During intercourse the tubular vaginal canal acts as a receptacle for the penis, secreting a watery substance which serves as a lubricant. Illustration 16 shows that the vaginal canal angles upward and backward from the vulva. In its normal state, the vagina is a soft and distensible organ, pliable enough to respond to the action of other muscles within the bony pelvis. During childbirth the vagina serves as the birth canal for the foetus after it leaves the uterus.

Illustration 17 shows other organs within the pelvis of the female. Notice the location of the bladder and the urinary duct, or urethra, which parallels the vaginal canal. The external urinary meatus is located immediately in front of the opening of the vagina.

● ● ●

THE FEMALE REPRODUCTIVE CYCLE

The mystery of reproduction is really the magic of body chemistry. The action of chemicals (called hormones) on the female reproductive system is responsible for bodily changes which make the reproductive organs receptive to conception each month and which permit the foetus to grow within the mother. Hormones also act to modify key reproductive organs during periods when the female does not conceive. This hormonal activity within the body thus periodically recycles the reproductive system, preparing it in the next month to be ready again for fertilization and pregnancy.

The hormones that control the female reproductive cycle are produced in various glands and carried by the bloodstream to organs of the reproductive system where they cause the activity of the organs to be modified. Illustration 18 shows the location of the glands that produce the major hormones involved in reproduction and the organs on which the hormones act. The arrows point from the glands to the affected organs.

As has already been stated, the female reproductive system usually releases only one egg every twenty-eight days. Upon release this egg is available for fertilization and the beginning of a new life. But during the entire month, the reproductive system is constantly undergoing subtle changes—preparing for ovulation, making the system receptive to fertilization, and then, if conception does not take place, readying the system to repeat the same cycle during the next twenty-eight days.

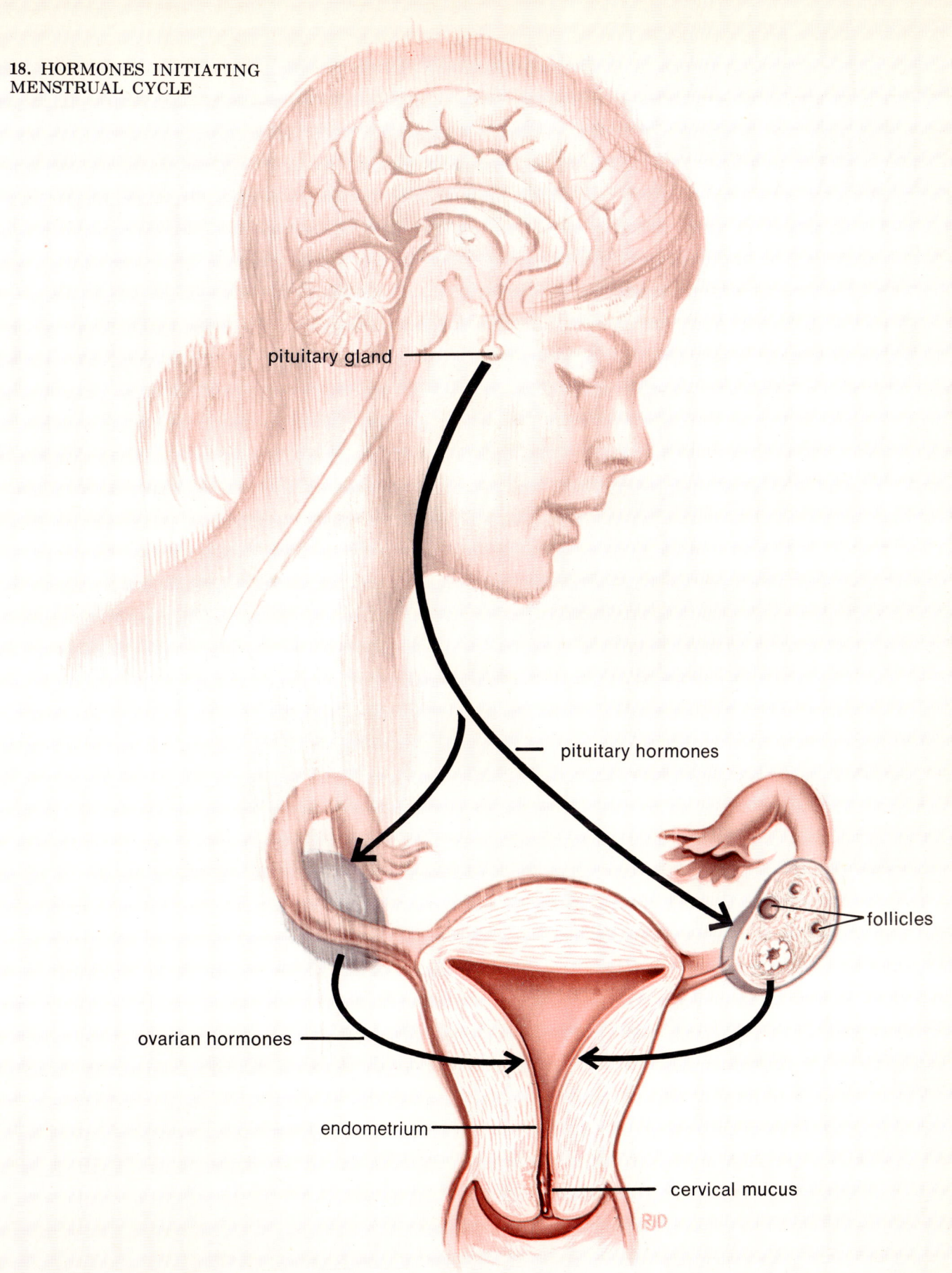

pituitary gland

pituitary hormones

follicles

ovarian hormones

endometrium

cervical mucus

For most women, the only evidence of this cycle of activity occurs during the menstrual period—during approximately four or five days each month when a small amount of blood is discharged from the vagina. Menstruation usually begins when the girl is about thirteen years of age and recurs monthly, except during pregnancy, until the age of about forty-five to fifty.

The reproductive cycle, or menstrual cycle, of the nonpregnant female is measured in days, beginning with the first day of menstruation and running to the first day of the next menstrual period. For most women, the cycle consists of twenty-eight days: days 1 to 5 encompass the period of menstruation. This time, plus the next eight days (up to day 13), is called the preovulatory phase. Ovulation usually occurs on the fourteenth day, and days 15 to 28 are called the postovulatory phase. These dates are approximate and may vary from woman to woman. Even in the same woman these dates can vary from month to month. Thus a normal cycle may range anywhere from twenty-one to thirty-five days.

The cycle begins when the hypothalamic area of the brain acts on a small gland at the brain's base. This gland, the pituitary, produces substances called gonadotrophic hormones. One of these hormones acts on the ovaries in two ways: (1) It stimulates the ovarian follicles, causing them, and the eggs within them, to grow. (2) It acts on the ovaries, causing them to produce another hormone called oestrogen. Oestrogen is one of the two important female sex hormones.

The oestrogen produced by the ovaries is then carried by the bloodstream to the uterus where it stimulates the growth of the lining wall of the uterus or womb. The tissue of this wall is called the uterine lining, or endometrium. The growth of the endometrium prepares the uterus to accept the egg in the event that fertilization takes place.

During the first seven days of the menstrual cycle, the action of one of the gonadotrophic hormones produced by the pituitary gland causes several follicles in both ovaries to develop. Usually about the seventh day, one follicle becomes dominant and continues to grow while the others regress. Occasionally, however, more than one follicle will mature and more than one egg will be released. This is one of the ways in which multiple births can occur. When twins are born as the result of the fertilization of two separate eggs, they are called fraternal twins. They will not be identical, since they did not develop from the same egg.

In Illustration 20, the largest follicle is the one that has

19. FEMALE REPRODUCTIVE CYCLE IN DIAGRAMMATIC FORM

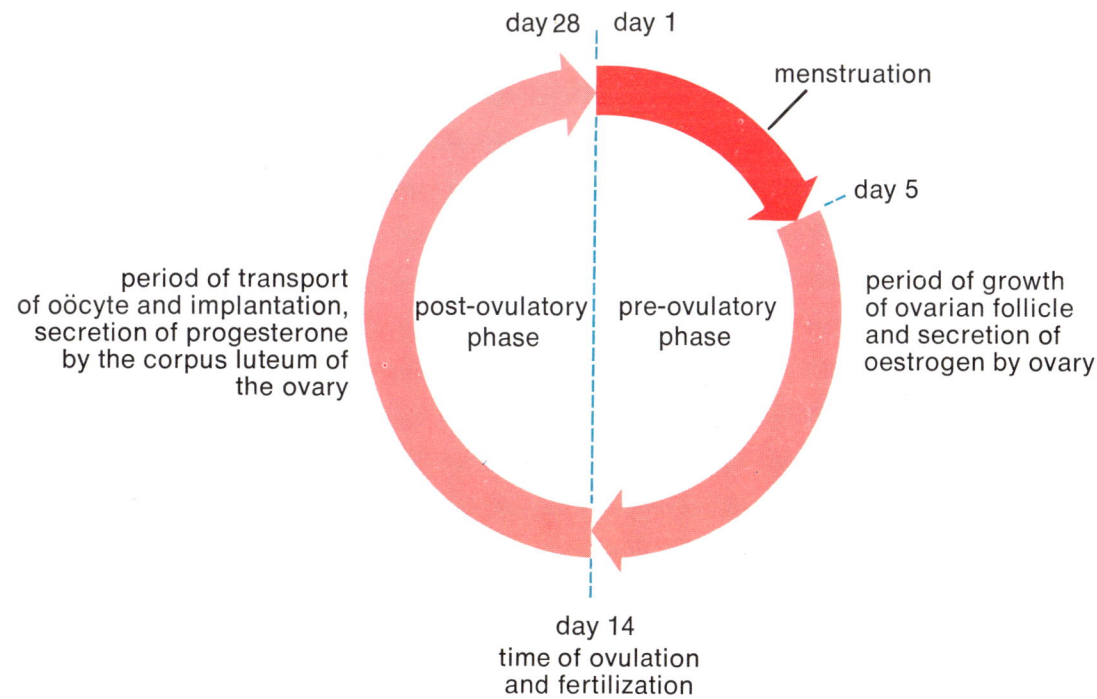

day 28   day 1

menstruation

day 5

period of transport
of oöcyte and implantation,
secretion of progesterone
by the corpus luteum of
the ovary

post-ovulatory
phase

pre-ovulatory
phase

period of growth
of ovarian follicle
and secretion of
oestrogen by ovary

day 14
time of ovulation
and fertilization

continued to develop, and is being prepared for ovulation. The other follicles have already started to regress. Also, the uterine lining is still thin and there is not much cervical mucus, a substance that helps transport the sperm and assure their survival.

On about the fourteenth day ovulation, or release of the egg, takes place. This occurs when the pituitary gland has secreted sufficient amounts of a second gonadotrophic hormone, called luteinizing hormone, to cause the ovarian follicle to rupture, releasing the egg so that it can be picked up by the open end of the adjacent fallopian tube.

As Illustration 20 shows, the action of the oestrogen on the uterus has by now caused the endometrium to thicken and become laced with many tiny blood vessels. Also shown is the clear mucus in the cervical opening. This mucus is most pronounced at the time of ovulation and helps in sperm transport and survival.

The growth and rupture of the ovarian follicle may occasionally cause a small amount of abdominal discomfort. This mild, mid-cycle discomfort is referred to as *mittelschmerz,* which is German for "middle pain."

After the ovarian follicle ruptures and releases the egg at the time of ovulation, the follicle becomes modified and is then called *corpus luteum* (Illustration 21 shows this change). The *corpus luteum* produces the second ovarian sex hormone, progesterone.

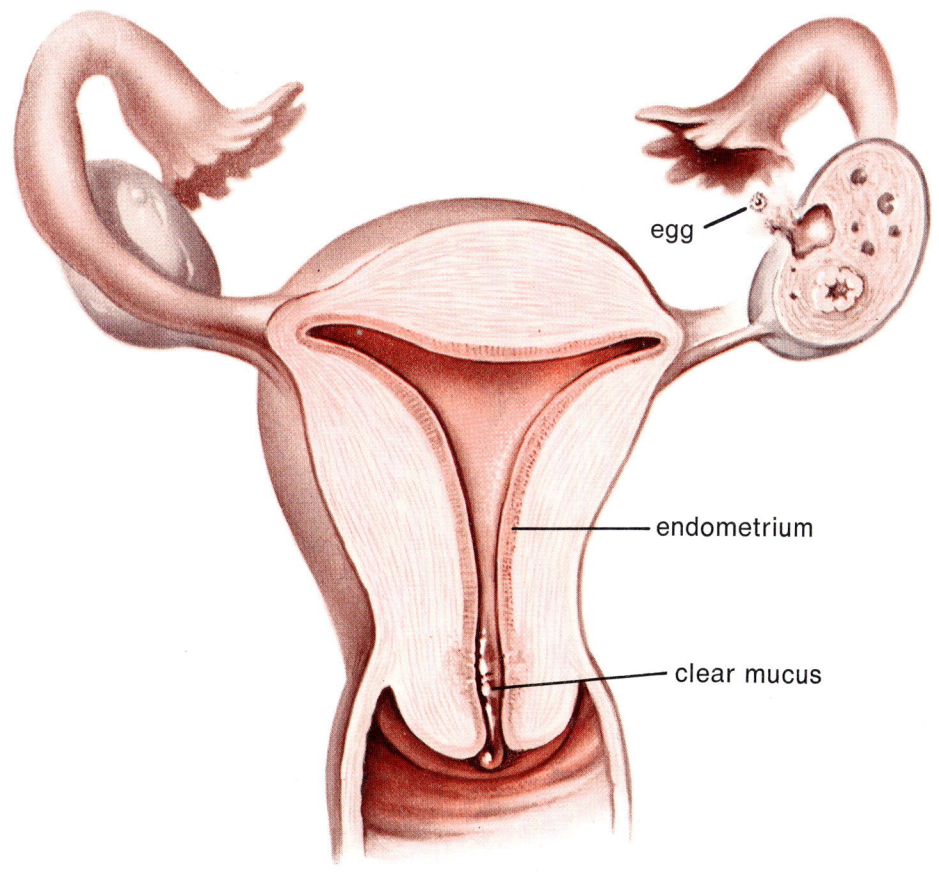

egg

endometrium

clear mucus

In the early phase of the female reproductive cycle, oestrogen is the dominant hormone, and the mucus at the cervix is clear, abundant, thin, and easy for the sperm to penetrate. In this postovulatory period, progesterone becomes the more dominant hormone, and the cervical mucus becomes thickened, cloudy, and scant. The two ovarian hormones, oestrogen and progesterone, also exert an effect on the vagina, the breasts, the glands of the skin, and on the amount of water held within the body tissue. The increase of water at this time causes some women to gain weight toward the middle or the end of the cycle.

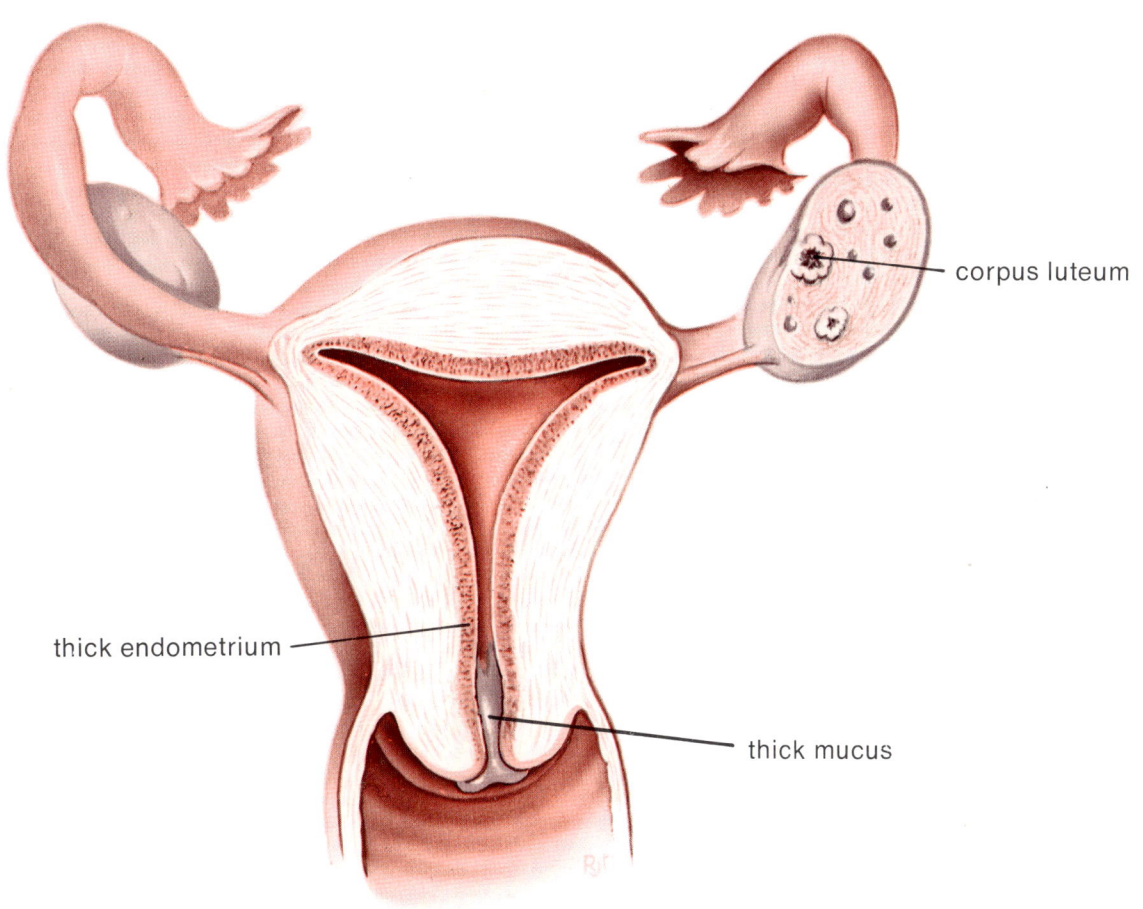

corpus luteum

thick endometrium

thick mucus

If, after the egg's release, fertilization does not occur the reproductive cycle ends with the onset of menstruation. This is the bloody discharge which results when the extra growth of the lining of the uterus is cast off. Menstruation marks the end of one reproductive cycle and the beginning of the next. It occurs when the ovaries decrease their production of hormones near the end of the twenty-eight-day period. When menstruation is accompanied by cramps or the sensation of heaviness in the pelvis, the woman is said to be suffering with dysmenorrhoea.

Illustration 22 shows the uterine lining falling away and passing out of the uterus through the cervicle canal. When the discharge reaches the vaginal canal, it can be absorbed by a vaginal tampon or a pad (usually about 1 or 2 ounces of blood are lost). The *corpus luteum*, which until this time had been producing progesterone, now becomes smaller and eventually becomes only a small, thin scar in the ovary. At this point, it will be called a *corpus albicans*, and it will have stopped producing progesterone. Already, however, the pituitary gland has resumed production of the gonadotrophic hormone that stimulates the development of other follicles, one of which will become dominant during the next reproductive cycle.

If fertilization does take place during the reproductive cycle, another hormone acts on the *corpus luteum* to keep it active and prevent the regression that leads to menstruation. When this happens, the endometrium of the uterus maintains its lush blood supply in order to nourish the embryo as it grows within the uterus.

● ● ●

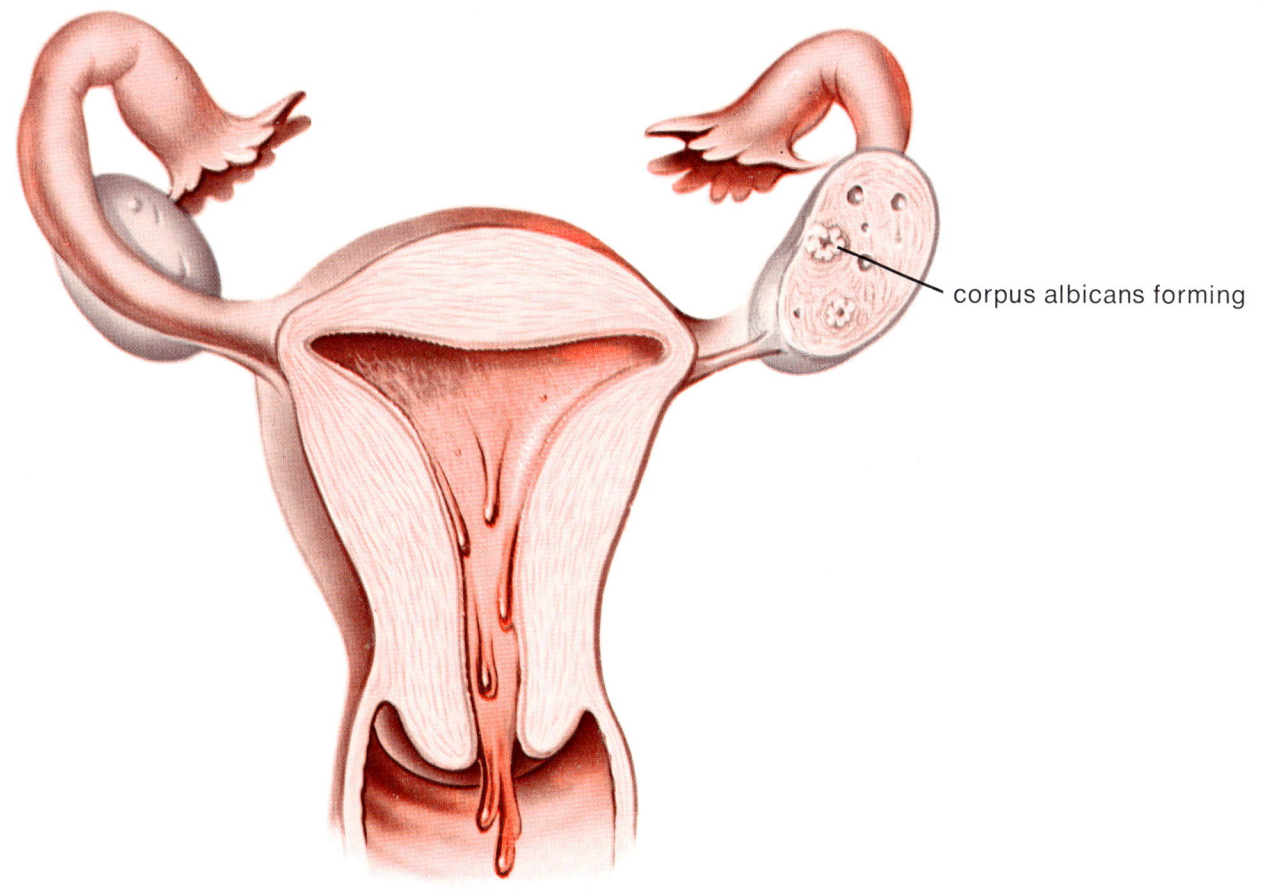

corpus albicans forming

FERTILIZATION AND IMPLANTATION
COITUS

Coitus is another name for sexual intercourse. It is the sex act which places the sperm of the male within the vagina, and against the cervical canal of the uterus.

Illustration 23 shows how the erect penis distends the vagina. The path that the spermatozoa follow from the testes through the *vas deferens* and through the urethra of the male genitourinary system is also shown. When ejaculation occurs during sexual intercourse, the semen containing the sperm is deposited at the upper end of the vagina at the cervix of the uterus. From this point, the sperm must travel through the cervical mucus, across the uterus, and up into the fallopian tube where they meet the egg and fertilize it.

Illustration 24 shows the position of the internal female reproductive organs during coitus. Highlighted is the area of the cervix and the upper vagina. When ejaculation occurs, the semen containing the sperm covers the cervix and pools in the indentation where the uterus protrudes into the upper part of the vagina.

At the completion of the sex act, the penis is withdrawn from the vagina, but most of the semen, containing several hundred million sperm, remains in the seminal pool in the vagina (see Illustration 24). Within a few minutes, millions of spermatozoa begin their journey through the female reproductive system, a journey that may take many hours and, sometimes, even several days. Obstructing the path of the sperm by mechanical or chemical methods will prevent conception. Techniques for accomplishing this are discussed in detail in the section on contraception.

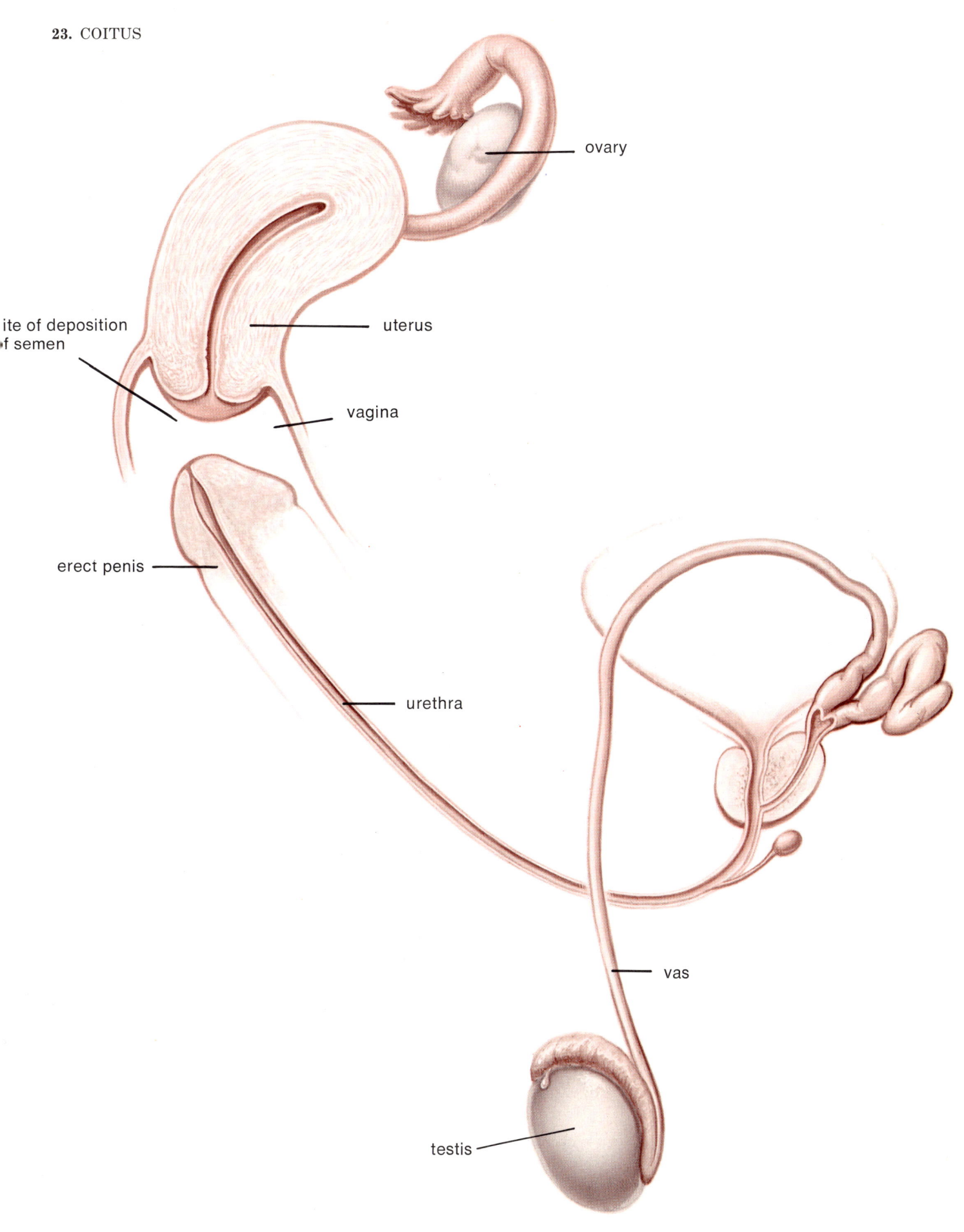

ovary

ite of deposition
f semen

uterus

vagina

erect penis

urethra

vas

testis

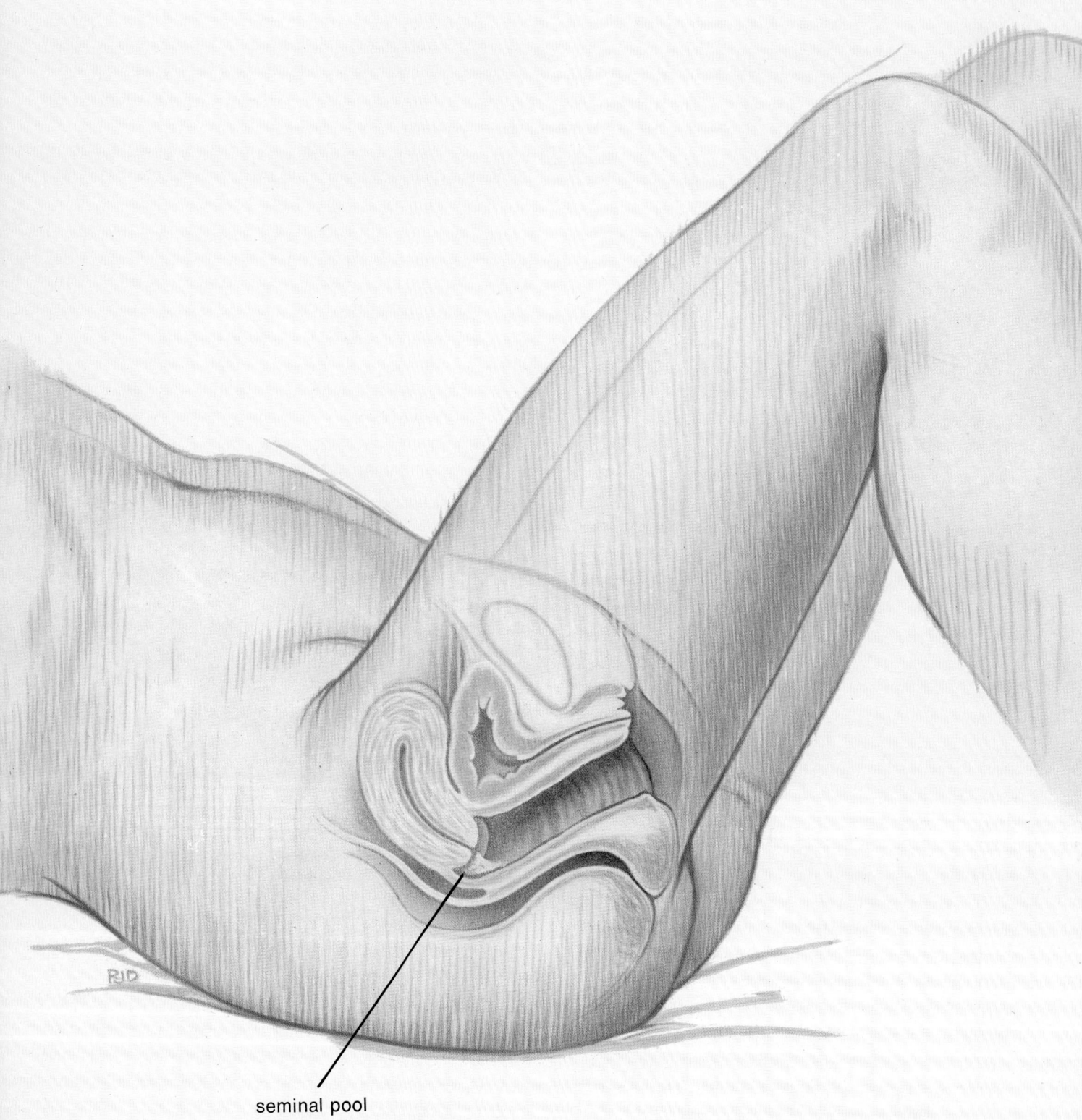

seminal pool

FERTILIZATION

The biological union that occurs when the sperm and the egg are united within the fallopian tube of the female is termed fertilization. A broader term, conception, refers to fertilization and the associated changes that take place within the body of the female which lead to the development of the fertilized egg in the uterus. Pregnancy begins with the implantation in the uterus of the fertilized egg.

As noted earlier, ovulation takes place at about the midpoint of the female reproductive cycle. The egg is released from the follicle of the ovary and almost immediately enters the large open end of the adjacent fallopian tube. A small amount of fluid accompanies the egg and carries it about one-third of the way down the tube. During the first few hours following ovulation the egg matures to become ready for fertilization. At this time all that is required is the presence of fertile sperm. The period during which the egg can be fertilized is less than twenty-four hours—a very short time in a twenty-eight-day reproductive cycle. But the motile, fertile sperm have a longer life span and can survive in the tube for even a day or two before the arrival of the egg. (This is an important point to remember if one practices the rhythm method of birth control.)

Illustration 26 shows the pathway the sperm take as they travel from the cervix to the point of fertilization within the tube. Sperm deposited at the cervix can survive in the mucus of the cervix for two days or longer, depending on the condition of the mucus. The longest period for sperm survival occurs in the immediate preovulatory phase of the reproductive cycle when the mucus is especially favourable to sperm penetration. Once deposited, the active sperm can move along the female reproductive tract at a rather rapid rate (as much as $\frac{1}{8}$ inch per minute), reaching the site of fertilization within a few hours.

Since sperm can survive for several days within the female reproductive tract and can also move rapidly from cervix to midpoint in the tube, fertilization can take place if coitus occurs anytime from several days *prior* to ovulation to twenty-four or more hours *after* ovulation.

25. OVULATION

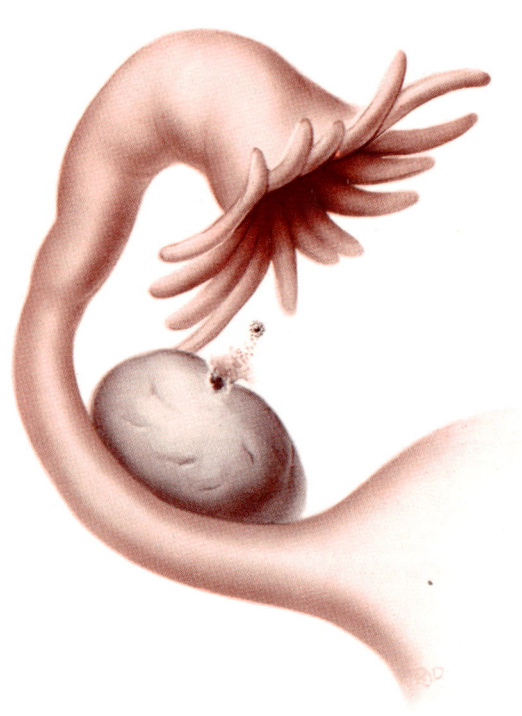

Although many millions of sperm are deposited against the cervix, only a few hundred or a few thousand reach the place within the tube where the egg is located. Scientists do not know why so many sperm are lost during the passage, but it has been established that the potency of the male (his ability to effect a pregnancy) is often directly related to the number of active, motile sperm he can release on ejaculation. Males with less than one hundred million active sperm per ejaculation often have difficulty effecting a pregnancy.

26. PATHWAY OF SPERM WITHIN THE FEMALE

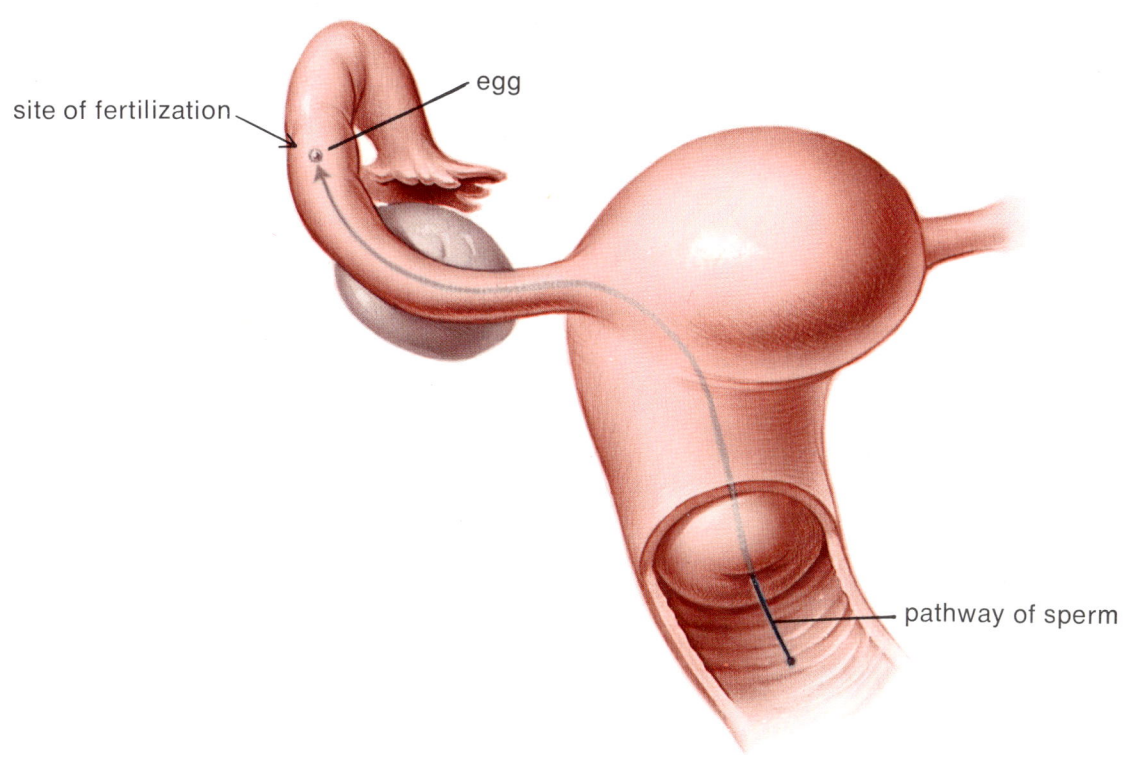

site of fertilization

egg

pathway of sperm

Illustration 27 shows the process of fertilization magnified several hundred times. Of the millions of sperm deposited, a few hundred have reached the egg, but only one is necessary for fertilization. Some scientists think that the accompanying sperm may be helpful in the process of fertilization, but this has not been definitely established.

In the illustration, the large, round central structure is the egg. The small round body is a polar body, which contains material not needed by the developing egg. Surrounding the egg is a less dense material, called the *zona pellucida*, through which the sperm must pass in order for fertilization to occur.

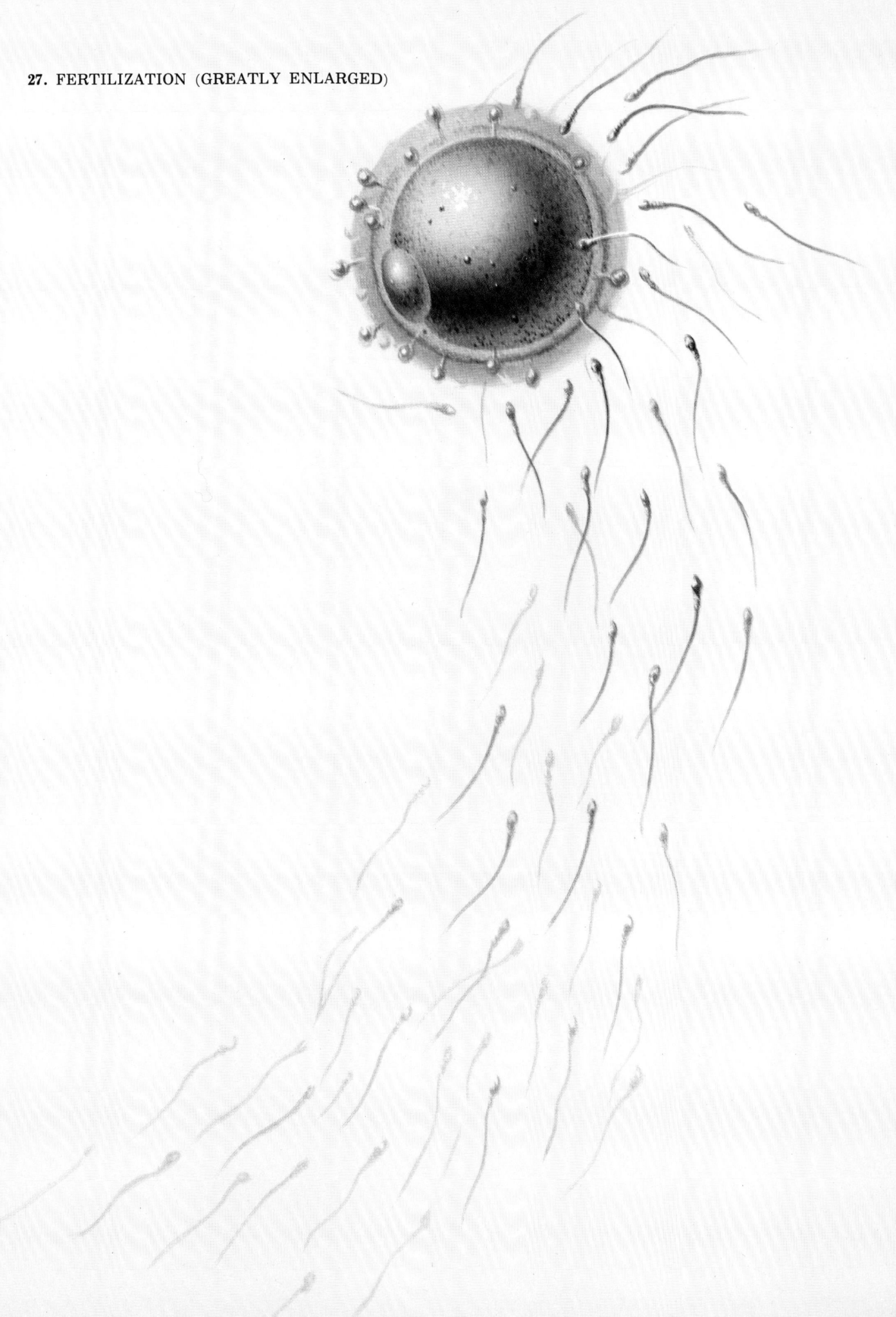

27. FERTILIZATION (GREATLY ENLARGED)

The inherited characteristics of the baby to be born as the result of fertilization are determined by material within the egg and sperm called chromosomes. Each mature sperm contains chromosomes which carry inheritable characteristics of the father. Similarly, each mature egg contains genetic material carrying the inheritable characteristics of the mother. When the egg and sperm are united, the chromosomes from the male and the female are also united, and the child receives inherited characteristics from both parents.

Once the sperm has penetrated the *zona pellucida* of the egg, and the nucleus of the sperm and egg have merged, the egg begins to divide. This cell division, or cleavage, usually starts within a few hours after fertilization. First the egg divides into two cells, then four, then eight—doubling the number of cells with each successive division. While this is happening, the egg is continuing its journey down the fallopian tube toward the uterus.

28. UNITING OF CHROMOSOMES FROM MALE AND FEMALE

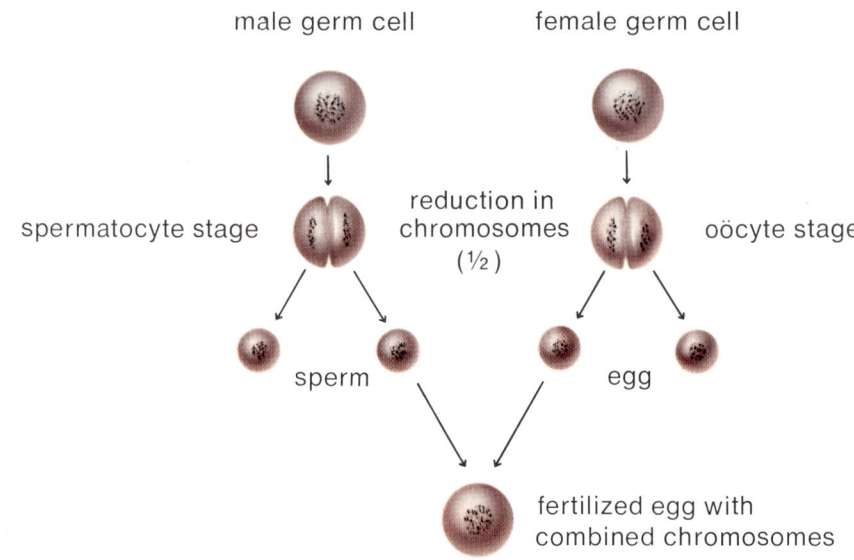

male germ cell   female germ cell

spermatocyte stage   reduction in chromosomes ($\frac{1}{2}$)   oöcyte stage

sperm   egg

fertilized egg with combined chromosomes

During this period, the fertilized egg becomes a ball-shaped cluster of rapidly dividing cells, called a morula, which under a microscope resembles a mulberry. The trip down the tube takes about four days, and, in this phase of growth, the morula utilizes nourishment that the egg has carried with it from the ovarian follicle and also receives nutrients from fluid available in the tube itself.

The pathway of the fertilized egg is shown in Illustration 29. By the time the egg reaches the uterus, it has become a hollow ball of cells and is now referred to as a blastocyst. Up until this time, the cells have been surrounded by the *zona pellucida*. Now the *zona pellucida* disappears and during the next two days, the blastocyst attaches itself to the lining of the uterus. This process of attachment is called implantation, and accomplishes pregnancy.

29. PATHWAY OF FERTILIZED EGG

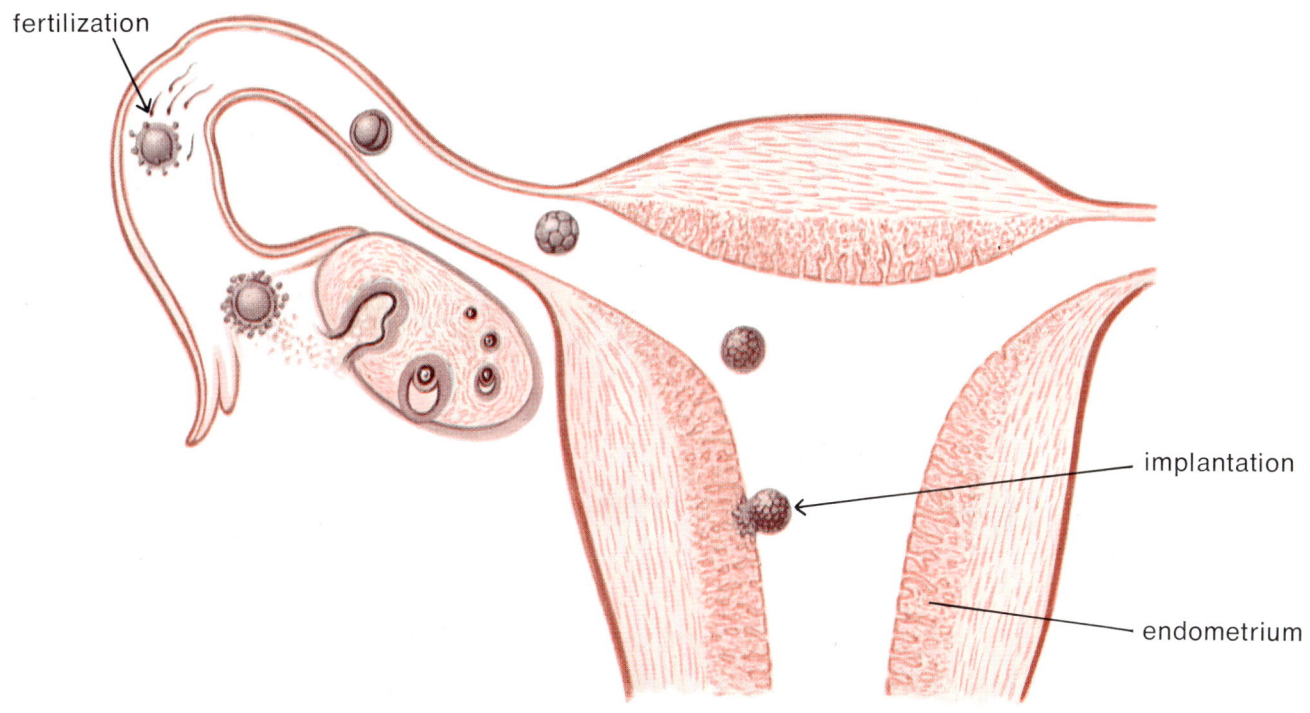

fertilization

implantation

endometrium

For implantation to occur, both the blastocyst and the lining of the uterus must be properly prepared. For example, if the fertilized egg has moved too rapidly down the tube and presents itself in the uterus before it has reached the blastocyst stage, implantation will not take place and pregnancy cannot be established. Also, if the endometrium, or uterine lining, is not ready to receive the blastocyst, the blastocyst will not adhere to the lining and, again, pregnancy cannot occur. When this occurs, the egg will simply degenerate.

Some contraceptives may act to speed the passage of the fertilized egg down the tube so that it arrives in the uterus too soon for implantation. Others may alter the development of the uterine lining by chemical or mechanical means so that it will not favour implantation. These methods will be discussed in the section on contraception.

Occasionally, as the fertilized egg starts down the fallopian tube, it divides into two cells and each cell separates and continues to develop independently. When this happens, *two* blastocysts arrive at the uterus. If both become implanted, a twin pregnancy will result. Since the twins will have come from the same egg and the same sperm, they will be identical. As indicated earlier, fraternal twins develop when *two separate eggs* are fertilized, and implant separately.

30. DEVELOPMENT AND ATTACHMENT OF FERTILIZED EGG (ENLARGED)

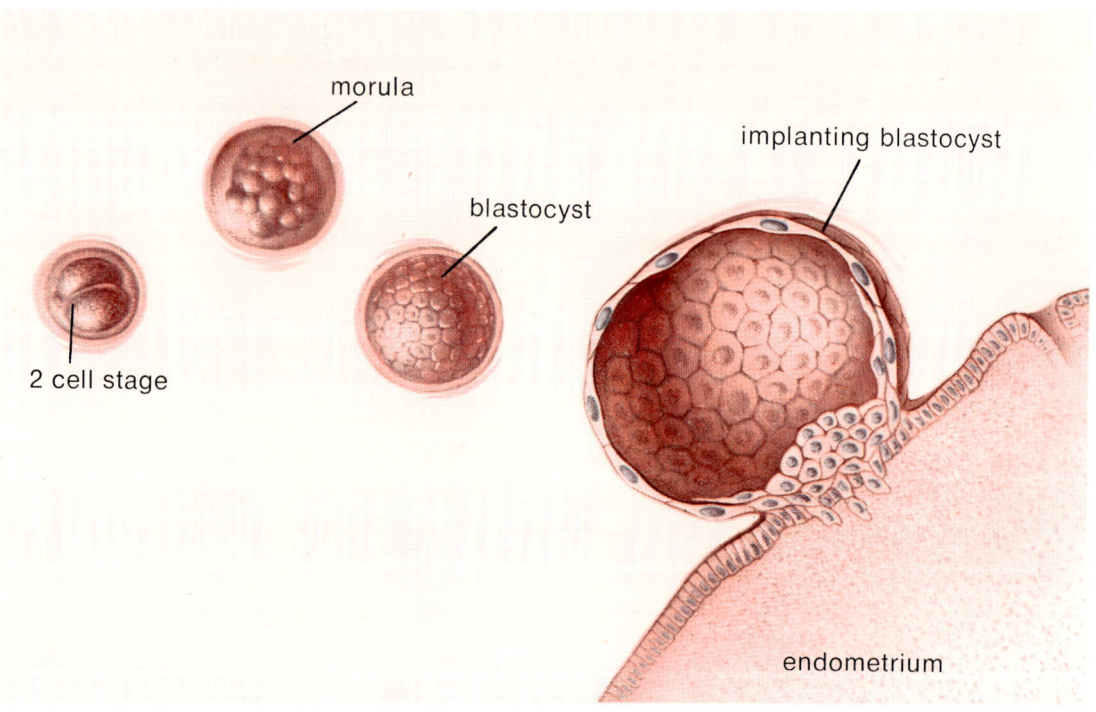

By the time the blastocyst is ready to attach itself to the endometrium of the uterus, it may contain as many as one hundred individual cells. These form a hollow sphere which usually implants in the lining of the upper portion of the uterus. A small number of the cells at one side of the hollow sphere will develop into the embryo. Most of the remaining cells form the trophoblast, and these cells develop into the placenta and associated membranes. The placenta—discussed in greater detail below—is a specialized organ that transfers nourishment from the mother to the foetus.

Implantation occurs on about the twentieth day of the cycle in which fertilization takes place. The blastocyst adheres to the uterine lining and rapidly penetrates into the tissue layers. Within a few days, the surface of the uterine lining seals over the blastocyst. To the naked eye, the implanted blastocyst appears as a slight bump on the surface of the uterus. On rare occasions, the blastocyst will implant along the fallopian tube or at other sites outside the uterus. When this happens, an abnormal form of pregnancy called an ectopic pregnancy results. Such pregnancies are usually terminated surgically since they are dangerous to the mother and provide no possibility of a live child.

The blastocyst, greatly magnified in Illustration 31, is entirely implanted within the uterine lining. The illustration shows it as it would appear on the twelfth day after fertilization (the twenty-sixth day of the reproductive cycle). The large cells at the centre will develop into the embryo, which will later become the foetus. The cells on the blastocyst surface form the trophoblast, which will develop into the placenta.

31. IMPLANTED BLASTOCYST, 12 DAYS AFTER FERTILIZATION (ENLARGED)

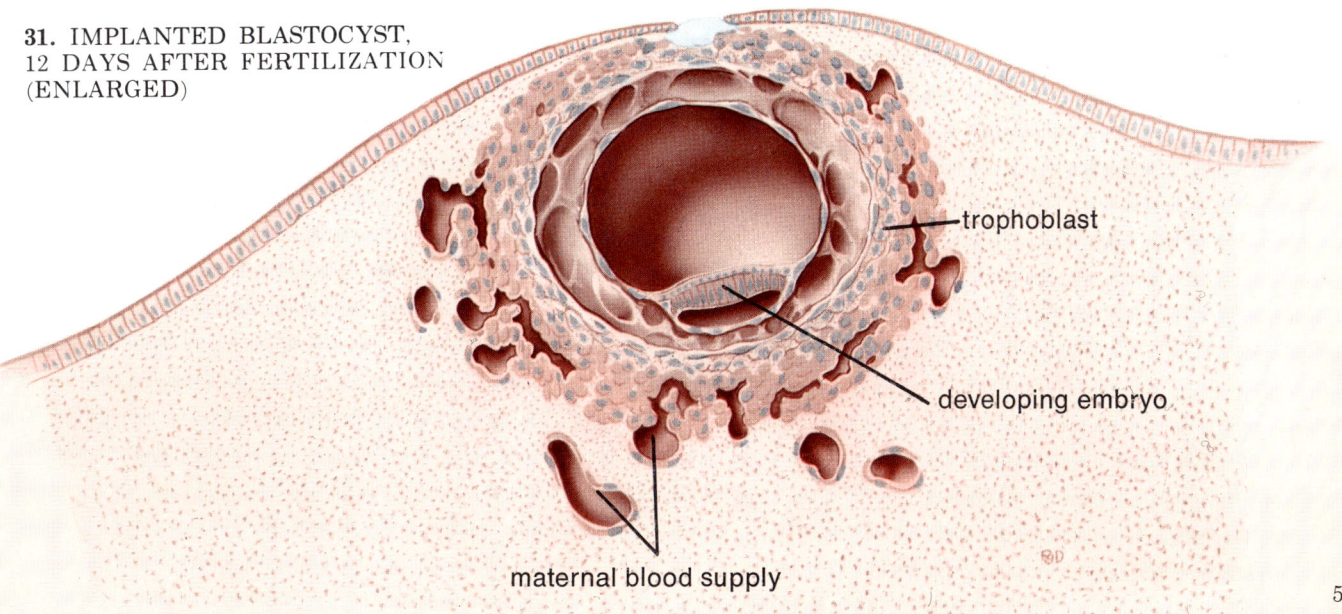

trophoblast

developing embryo

maternal blood supply

The placenta secretes a hormone called chorionic gonadotrophin, which acts to maintain the *corpus luteum* and its output of sex hormones. Since hormone production does not stop, the uterine lining remains healthy. (Menstruation occurs when this lining breaks down and is discharged through the vaginal canal.) The secretion of chorionic gonadotrophin continues to stimulate the ovaries during the early part of the pregnancy. The female sex hormones, oestrogen and progesterone, produced by the ovaries and the placenta act to prevent the release of additional eggs from the ovaries during the time the female is pregnant.

The presence of chorionic gonadotrophin in the female's blood and urine also serves as early evidence to the physician that a pregnancy has occurred. Since a urine sample from a pregnant woman about one week following a missed menstrual period will contain chorionic gonadotrophin, the urine test can be used to determine whether or not a woman is pregnant.

Illustration 32 diagrammatically presents the time sequence from ovulation through fertilization, tubal transport, attachment, and implantation of the blastocyst.

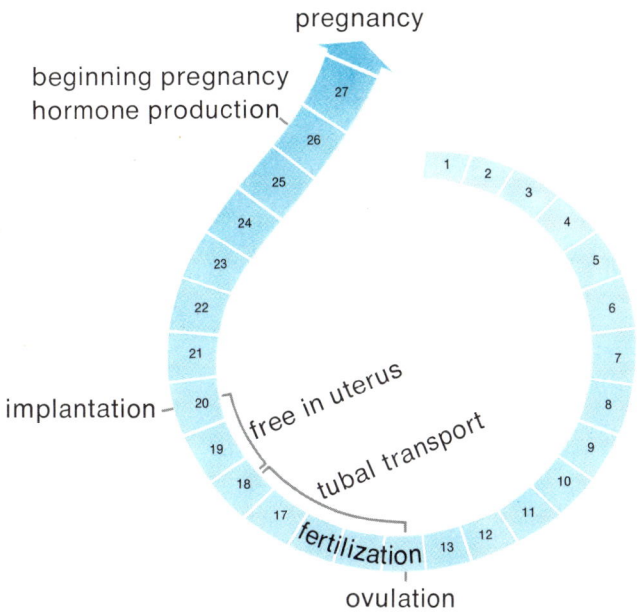

Illustration 33 shows drawings taken from seventeenth-century manuscripts dealing with human reproduction. On the left is a male sperm, which is pictured as containing a miniature individual, or homunculus; on the right is an opened uterus containing a miniature foetus. Fertilization and embryonic development were poorly understood until it was appreciated that both the male and the female made equal genetic contributions to the new individual.

33. 17TH-CENTURY REPRESENTATIONS OF HUMAN SPERM AND EARLY HUMAN CONCEPTUS

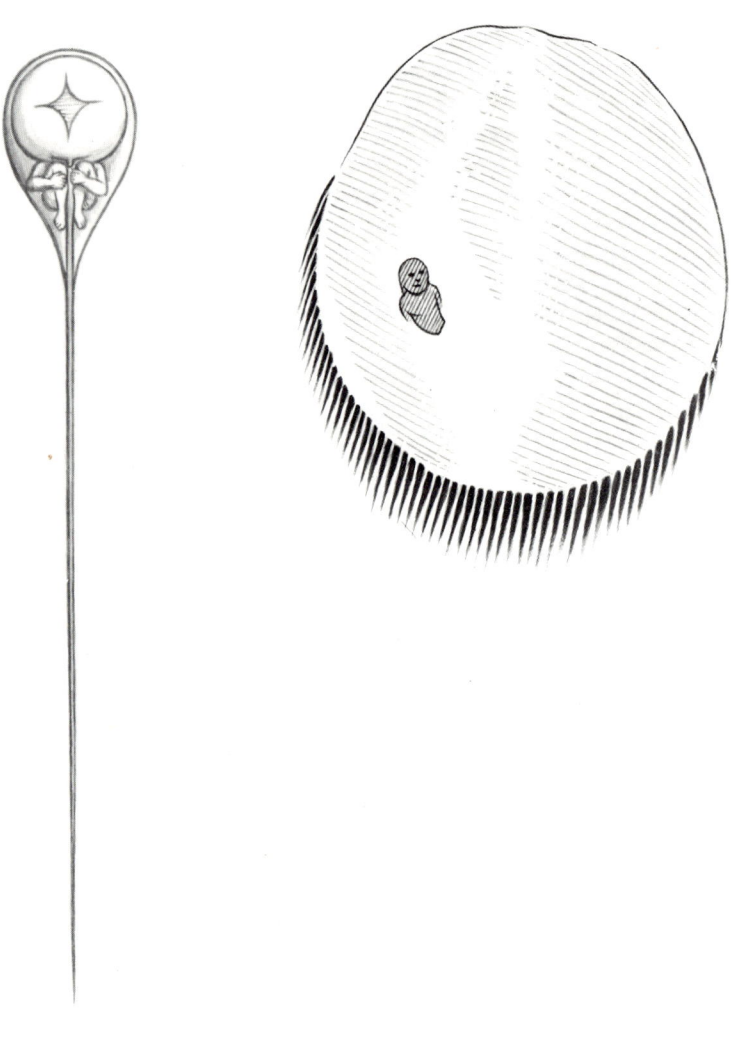

DEVELOPMENT AND GROWTH OF THE FOETUS

Doctors have two ways of calculating the day on which the baby will be born. Since fertilization occurs on the fourteenth day of the menstrual cycle and the delivery date is about 266 days after fertilization, doctors add 14 and 266 and figure that the date of birth will be 280 days (forty weeks or approximately nine calendar months) after the first day of the last menstrual period. Doctors also use this formula: First day of last menstrual period + 7 days − 3 months + 1 year. If the first day of the last menstrual period was June 1, the delivery date would be: June 1 + 7 days = June 8 − 3 months = March 8 + one year. Therefore, the child would be expected on March 8 of the following year.

The growth of the baby within the uterus takes nine calendar months. In talking about the development of the new life, doctors divide the nine months into 3 three-month periods called trimesters. The nine months may also be divided into the period of the embryo (approximately the first fourteen weeks) and the period of the foetus (approximately the last twenty-six).

It should also be noted that for medical purposes pregnancy is dated, not from the time of fertilization, but from the time of the woman's last menstrual period.

THE EMBRYONIC PERIOD

In the early stages of its development, the growth within the uterus is called an embryo. During this period, the basic organ systems are forming and the new life is very fragile. Anything that adversely affects the developing baby during this phase of growth may result in physical abnormalities such as cleft lip or the presence of extra fingers or toes. If the abnormality is severe, the embryo may die and be expelled from the uterus. During the first five months of pregnancy, the casting off of the embryo, or small foetus, is termed an abortion. About 10 per cent of all embryos are aborted spontaneously for one reason or another, the majority of these abortions occurring within the first few weeks of pregnancy.

Illustration 34 shows how the inside of the uterus would appear if it were inspected shortly after the missed menstrual period (about the fifth week of pregnancy). The small bump, shown in this magnified view, on the surface of the uterine endometrium is the implanted blastocyst. This is an external view, but the blastocyst is at the same stage of implantation as shown in the cross-section drawing in Illustration 31. The little holes are the numerous pores of small glands present within the uterine lining.

Illustration 35 shows the human embryo within the uterus at the sixth week of pregnancy (the fourth week after the egg has been fertilized).

34. MAGNIFIED VIEW OF IMPLANTED BLASTOCYST

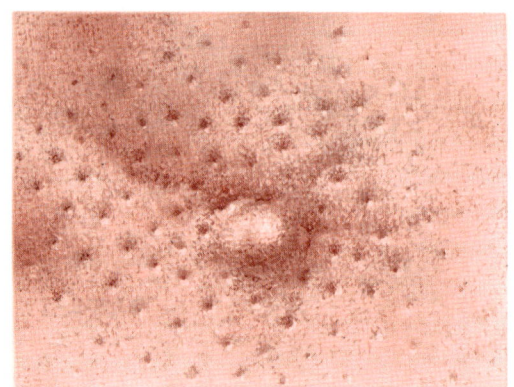

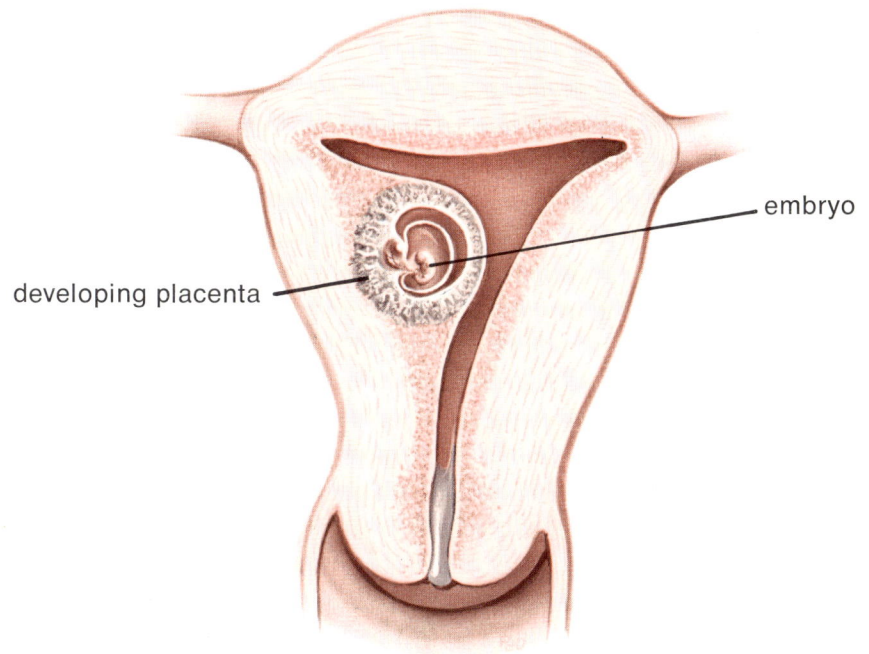

embryo

developing placenta

Illustration 36 shows the embryo in the tenth week of the pregnancy. By now, the embryo lies within a fluid-filled sac called the chorionic vesicle. This sac will later develop into the placenta and will contribute to the development of the membranes that surround the baby during the period it remains within the uterus. A body stalk connects the embryo to the chorionic vesicle. Through this stalk the embryo is able to receive nourishment from the rich blood supply in the endometrium (the lining of the uterus) where it is in contact with the chorionic vesicle. Later, the body stalk will develop into the umbilical cord.

The illustration shows that the head of the embryo is large in relation to the body. The toes and fingers are just forming, and the eyelids have not yet developed. At the end of another four weeks' development, all human features will be recognizable.

● ● ●

THE FOETAL PERIOD

By the fourteenth week, the embryo is called a foetus. The foetal period extends from this time until the birth of the infant. During this period, the foetus becomes more and more human-looking. The head now grows more slowly than the body, the arms and legs lengthen, fingernails begin to appear, and external reproductive organ develop. If it were possible to examine the foetus at this time, the sex of the unborn infant could be determined. The sex of the foetus could also be determined by examining under a microscope cells which are present in the amniotic fluid surrounding the foetus. To do this, however, the doctor would have to surgically puncture the amniotic cavity. This procedure is not undertaken except under very unusual circumstances having to do with the baby's health. So, until the baby is delivered, no one knows whether it is a boy or a girl.

Illustration 37 shows pregnancy in the fourteenth week. The uterus has begun to increase in size in order to accommodate the growing foetus, and the external appearance of the mother begins to evidence her pregnancy. Her abdomen has begun to expand and her breasts have begun to enlarge. The foetus itself now measures approximately 3 inches from head to buttock. The head is still large compared with the body but the foetus is beginning to have the appearance of a human being. A fluid-filled sac, called the amnion, surrounds and helps protect the foetus. In turn, the amnion is encased in an outer membrane called the chorionic sac, or chorion. At this stage of development the chorion and amnion make up the foetal membranes.

36. HUMAN EMBRYO—10 WEEKS OF PREGNANCY (ACTUAL SIZE)

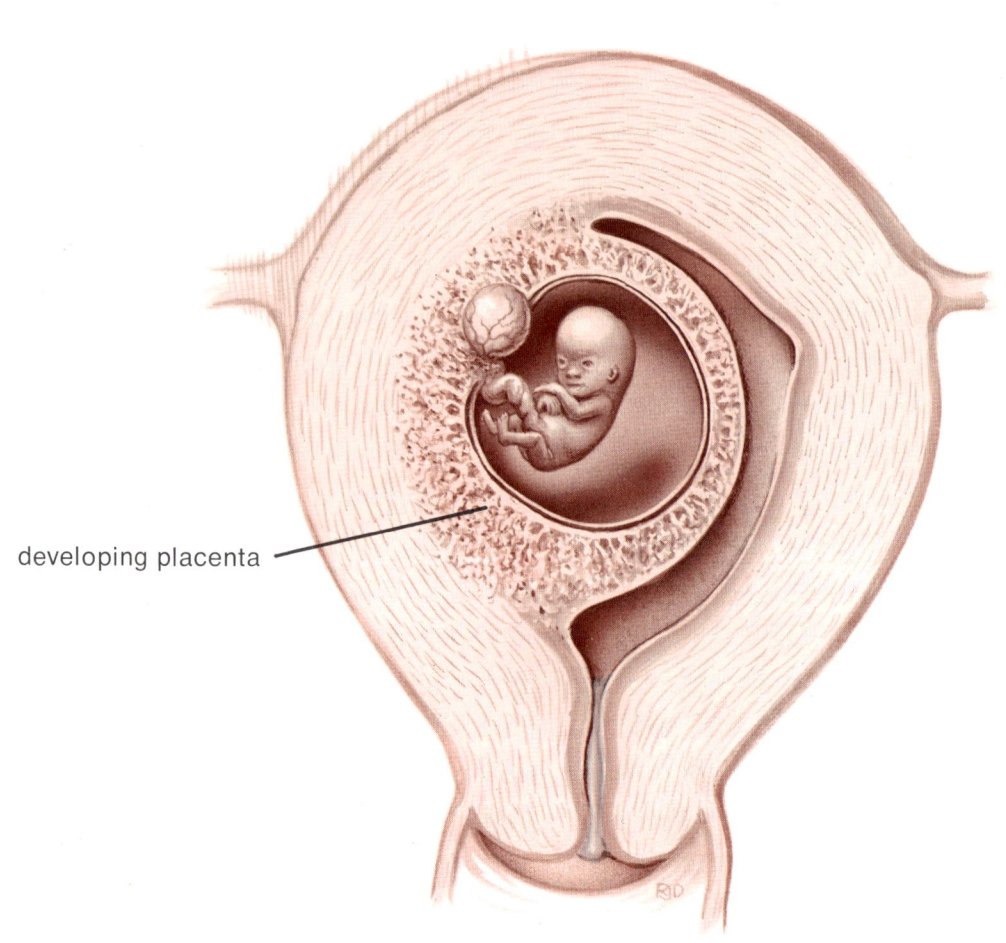

developing placenta

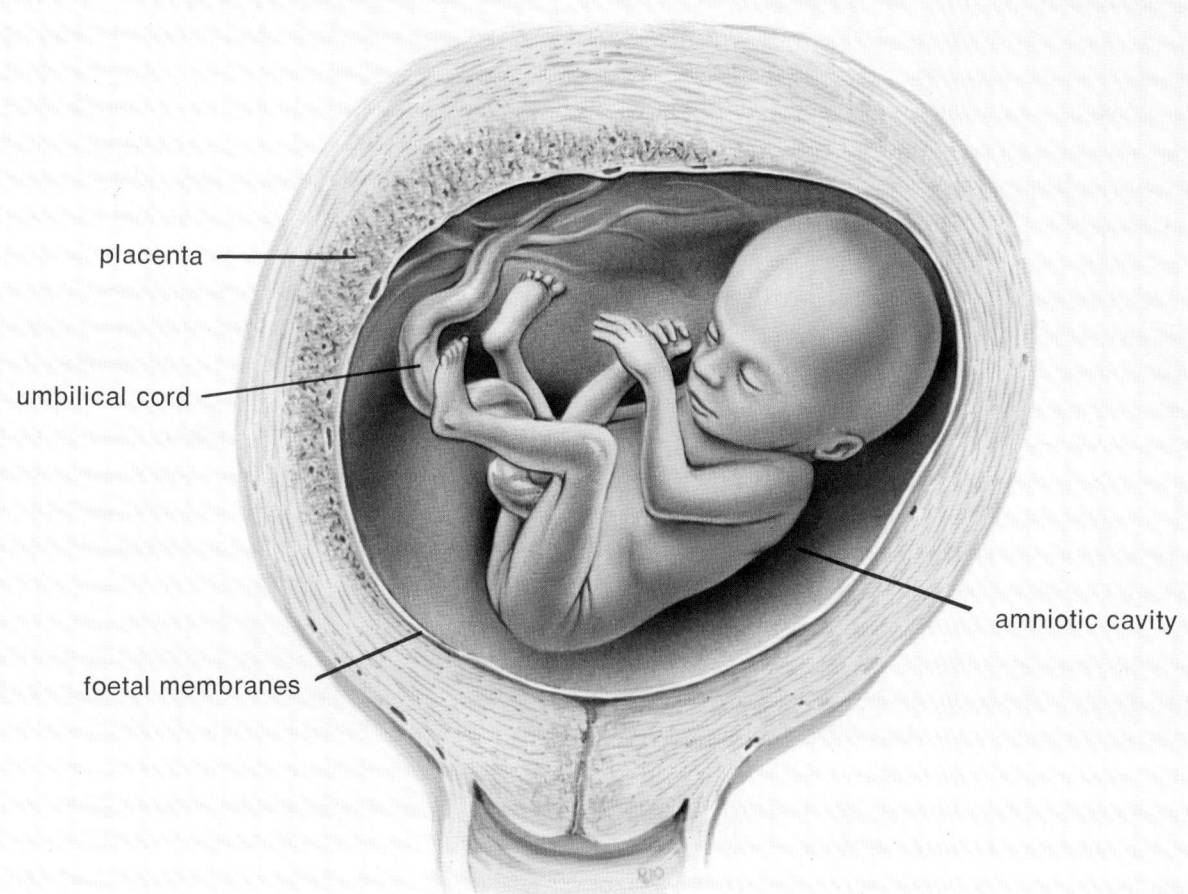

placenta

umbilical cord

foetal membranes

amniotic cavity

Also shown in Illustration 37 are the placenta and the umbilical cord, which by this time have assumed their mature form. The foetus is connected to the placenta by the umbilical cord. Within the umbilical cord are blood vessels which carry the blood between the placenta and the foetus. During the nine months of growth, the placenta plays an important role in providing nutrition for the foetus. Attached to the uterine wall, the placenta permits the bloodstreams of mother and foetus to be very close to each other, although these two bloodstreams never touch, and the blood in each remains separate. Oxygen and nutrients pass through the placenta, from the bloodstream of the mother to the foetus, and waste products from the foetus pass in the reverse direction into the bloodstream of the mother.

The placenta also continues to secrete the hormone chorionic gonadotrophin and to produce the hormones oestrogen and progesterone. These hormones act on the hypothalamus and the pituitary to block the secretion of the pituitary gonadotrophic hormones and thus prevent the ovaries from releasing additional eggs during pregnancy. (One method of contraception is to give the nonpregnant female drugs which contain artificially made hormones similar to oestrogen and progesterone. These drugs act exactly like the naturally occurring hormones in their ability to block ovulation. By inhibiting ovulation, these agents prevent conception.)

The placenta also produces another hormone, placental lactogen, which probably acts to prepare the breasts to produce milk.

The middle three months of pregnancy are referred to as the mid-trimester, and it is during this period that the signs of pregnancy become quite clear. By about the twentieth week, a doctor examining the mother can hear the foetal heartbeat and, if an X-ray is taken, he can see the foetal skeleton. The mother can feel the movement of the foetus as its arms and legs make contact with the foetal membranes, which by now are pressed against the walls of the uterus. At first the movement is only a slight flutter, but later it becomes much more pronounced. During the mid-trimester, there is still room for the foetus to turn about freely within the fluid-filled amniotic cavity.

If for any reason the foetus is delivered during this trimester, it can breathe, cry, and move, but because of its very small size, has little chance of surviving.

Illustration 38 shows the pregnancy at twenty-four weeks. The overall length of the foetus is now 12 inches and the head-to-buttocks length is 8 inches. The foetus has grown rapidly, now weighs approximately 1½ pounds, and has assumed its normal, head-down position. The placenta and the membranes that surround the foetus can be readily seen, but they now occupy only one-third of the total mass—much less than in the first trimester. The uterus has expanded noticeably to accommodate the growing foetus.

During the third, and last, trimester of the pregnancy the foetus increases rapidly in weight and size. About 50 per cent of its weight is added during its last 2½ months in the uterus. By the thirty-eighth week of pregnancy, the foetus is called a term foetus because birth at this time will usually result in the delivery of a normal-sized baby. A term foetus usually weighs approximately 7 pounds and measures about 20 inches in overall length. A foetus weighing less than 5 pounds at birth is called premature.

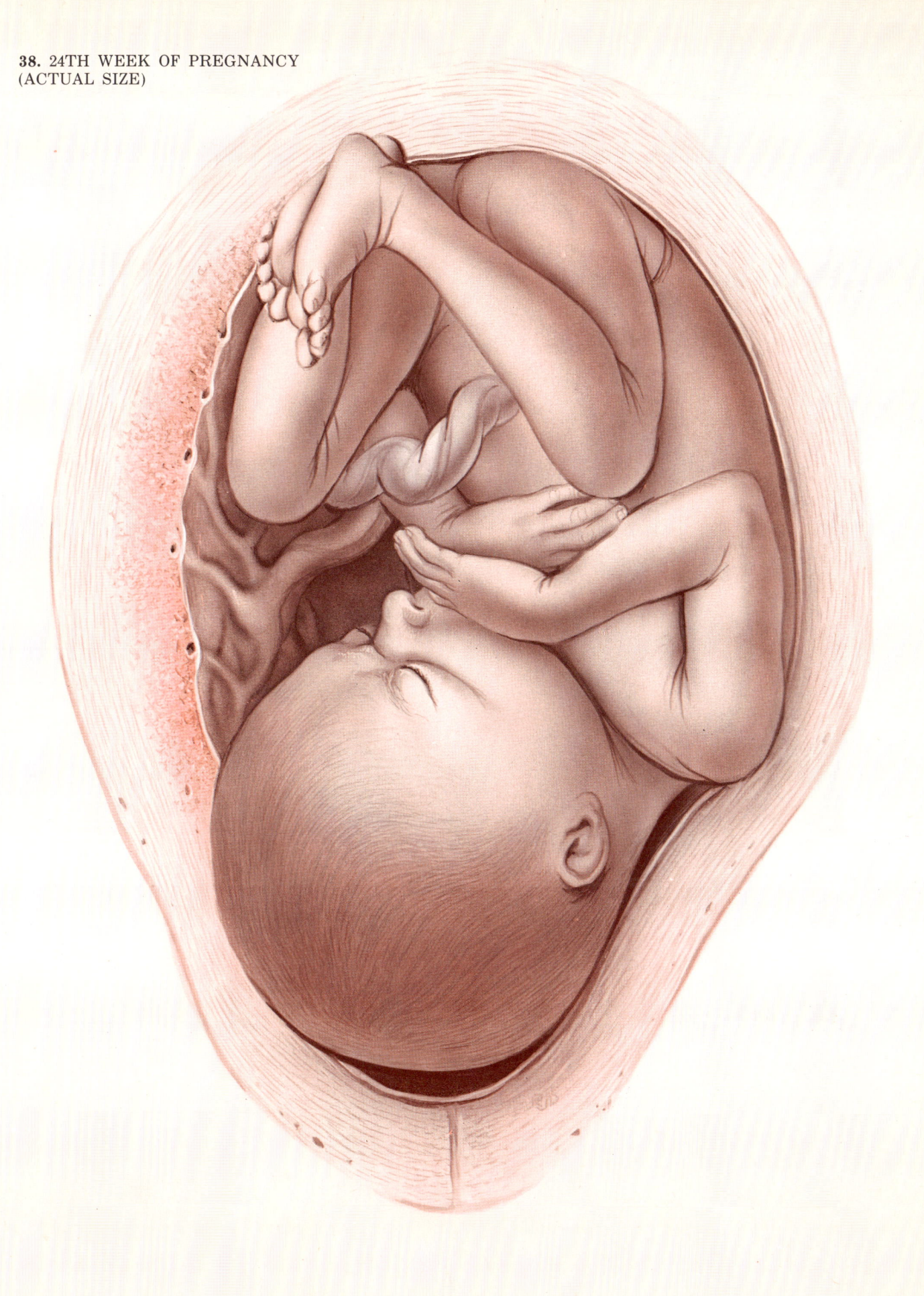

Illustration 39 shows a term foetus, actual size. The head is the largest part. Its size is important, since it must pass through the birth canal at the time of delivery. If the birth canal is too small to allow normal passage, the doctor may perform an operation called a caesarean section and remove the baby through an opening in the mother's uterine wall and abdomen.

The term foetus is rounded as the result of fat that has been deposited on its body. The head is covered with quite a bit of hair, and the fingernails extend beyond the tips of the fingers. The skin is covered with a whitish, fatty material called vernix caseosa, which comes from the secretion of glands within the skin and is most prevalent in the folds of the skin. The foetus now fills the uterus; its head is well down in the pelvic area of the mother—the ideal position for it when the birth process begins.

The foetus is now much larger than the placenta and weighs about eight times as much.

● ● ●

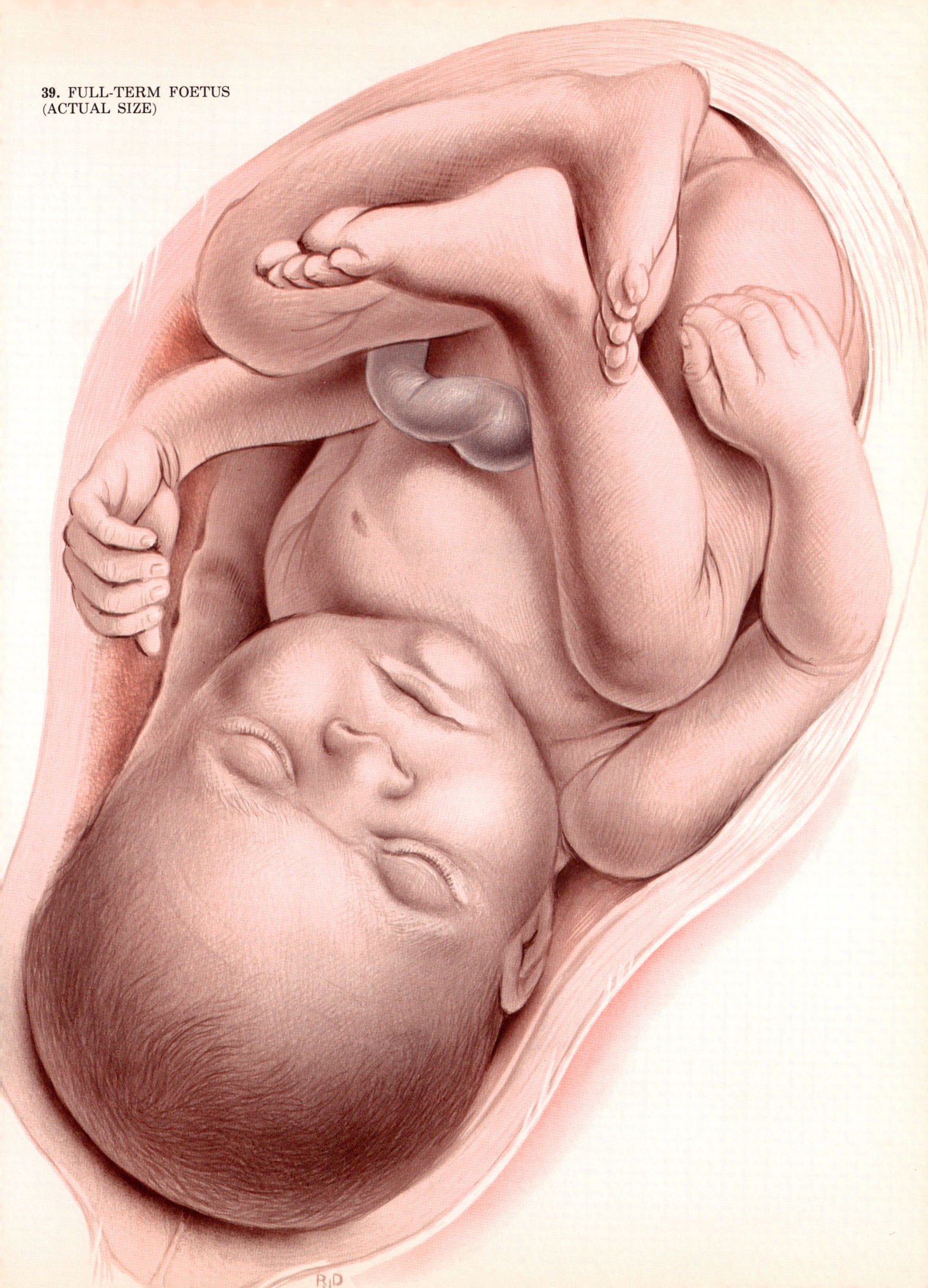

39. FULL-TERM FOETUS (ACTUAL SIZE)

BIRTH

LABOUR AND DELIVERY

STAGES 1, 2, AND 3

The new life that has been growing within the mother for approximately nine months is now ready to enter the world. The passage of the baby through the birth canal is called delivery, and the process by which this is accomplished by the mother is called labour. Delivery usually occurs between the thirty-eighth and fortieth weeks of pregnancy. Occasionally, a baby is born before thirty-eight weeks have passed. When this happens, the birth may be premature, and the baby's chance of survival depends upon its weight and state of development at birth.

When it is time for delivery, the mother begins to experience uterine contractions. This activity is considered the onset of labour. Each contraction is felt by the mother in her back and around to the front of her abdomen. At first the contractions occur at infrequent intervals and are not very intense. Later they become more intense and more frequent, eventually setting a rhythmic pattern. As labour progresses, the cervix becomes dilated and the foetus finally descends through the birth canal of the vagina and is delivered.

Not all the factors that control the onset of labour and delivery are completely understood. Doctors are sure, however, that hormones play an important role. Sometimes in order to initiate labour when the baby is due the doctor may administer a hormone called oxytocin (produced by the pituitary gland). Since the hormone oxytocin will often produce labour, it is likely that this hormone also plays an important role in natural labour.

Doctors divide the birth process into three phases. These stages are shown in Illustrations 40 to 45. The first stage of labour, shown in Illustration 40, begins with the dilation of the cervix and lasts until the cervix has dilated sufficiently to allow the foetal head to pass down into the vaginal canal. This stage is the longest period of the labour, and in the female having her first child may last as long as twelve hours. For mothers who have previously had children, the first stage seldom lasts longer than eight hours.

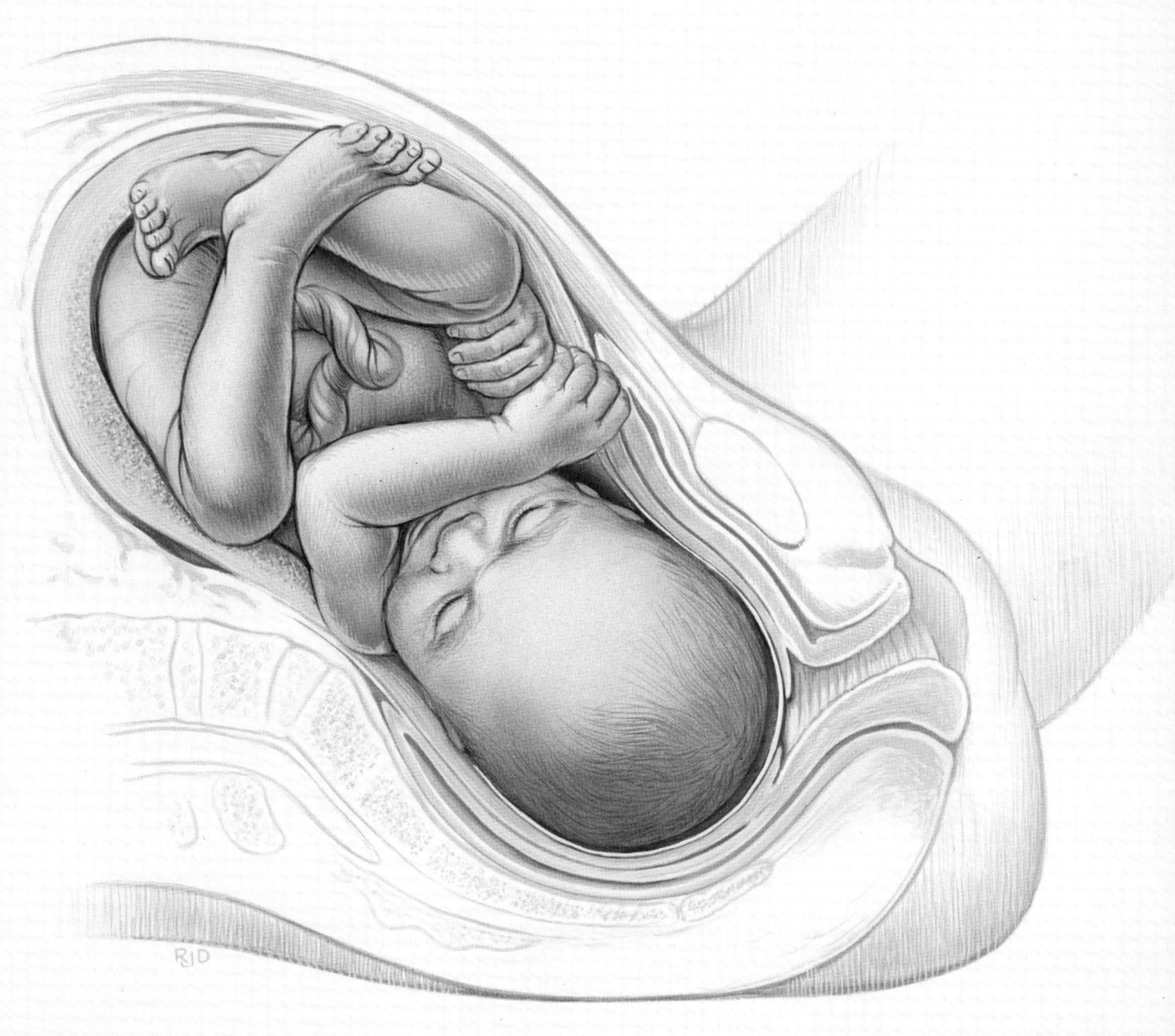

The second stage of labour consists of the foetus passing through the birth canal. This stage terminates with the actual birth of the baby and the cutting of the umbilical cord. Illustration 41 shows the beginning of the second stage of labour.

As the baby descends through the birth canal it usually rotates to allow the foetal head to pass more easily through the opening in the mother's bony pelvis. In Illustration 41 the foetal head is shown in the process of rotation, and the foetal membranes, sometimes called the "bag of waters," are shown rupturing over the head. At this time in the delivery, these membranes break and a large amount of amniotic fluid is released. Sometimes, the doctor will deliberately rupture the foetal membranes at an earlier stage of labour when he examines the mother.

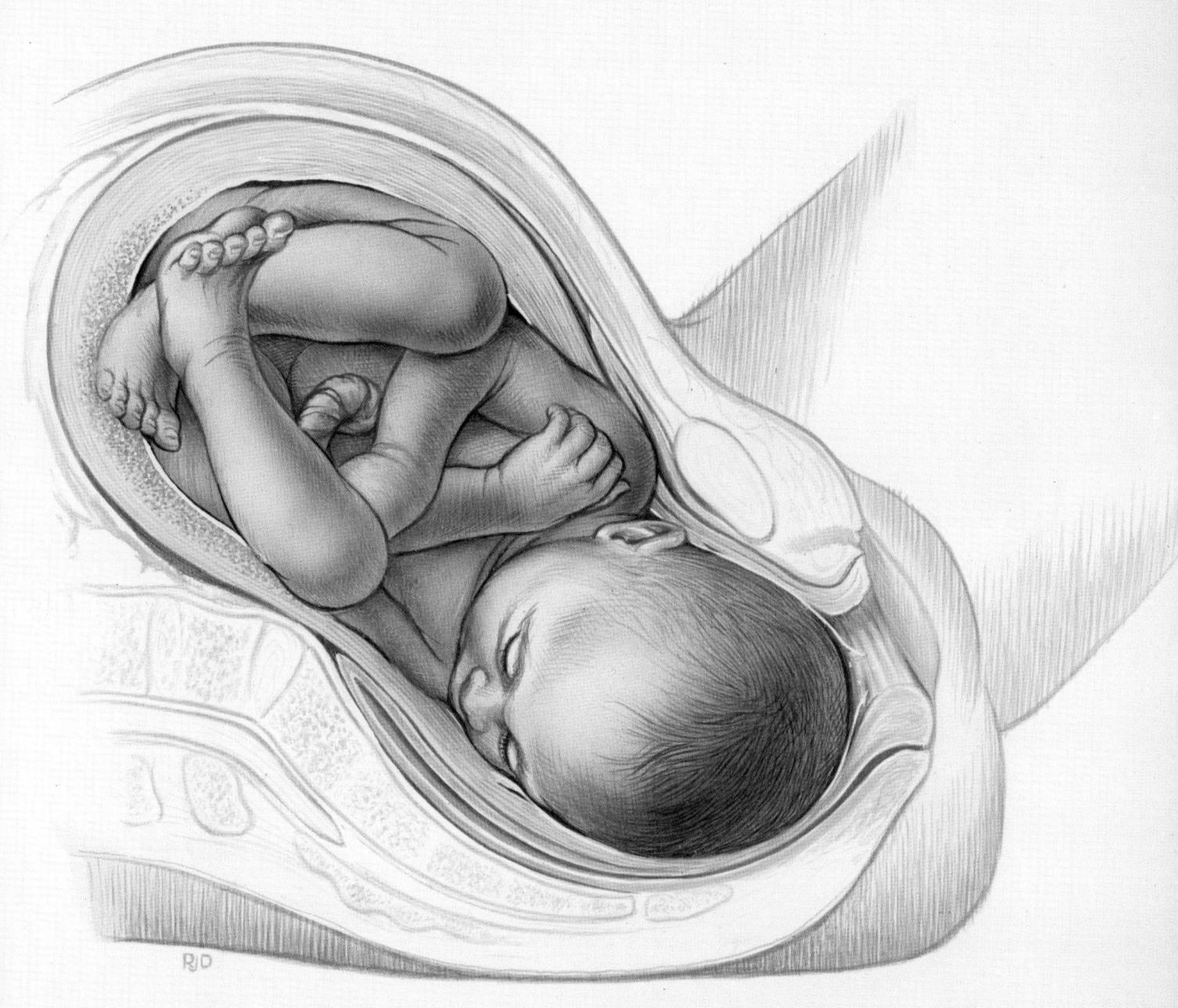

The second stage of labour may last only a few minutes if the mother has had children before. If it is her first child, this phase may last as long as two hours. Most of this time is required for the foetal head to traverse the birth canal. It is important during the second stage of labour that the mother voluntarily helps to push the foetus down through the birth canal. She does this by contracting the muscles in her lower abdomen and trying to expel the foetus. At the same time, the powerful uterine muscle mass presses against the foetal buttocks and extremities with each contraction of the uterus.

The thickness of the uterine muscle mass can be seen in Illustration 42. In this illustration, the foetal head has descended through the cervical opening into the vaginal canal. The canal has been greatly extended to permit passage of the head and shoulders. The vaginal opening will distend even more to permit final passage of the baby out of the birth canal. Sometimes the doctor will cut the vulva slightly to prevent undue stretching or tearing of the maternal tissues. Such an incision or surgical cut is called an episiotomy. After delivery, the doctor places a few surgical stitches in this small cut, which heals promptly.

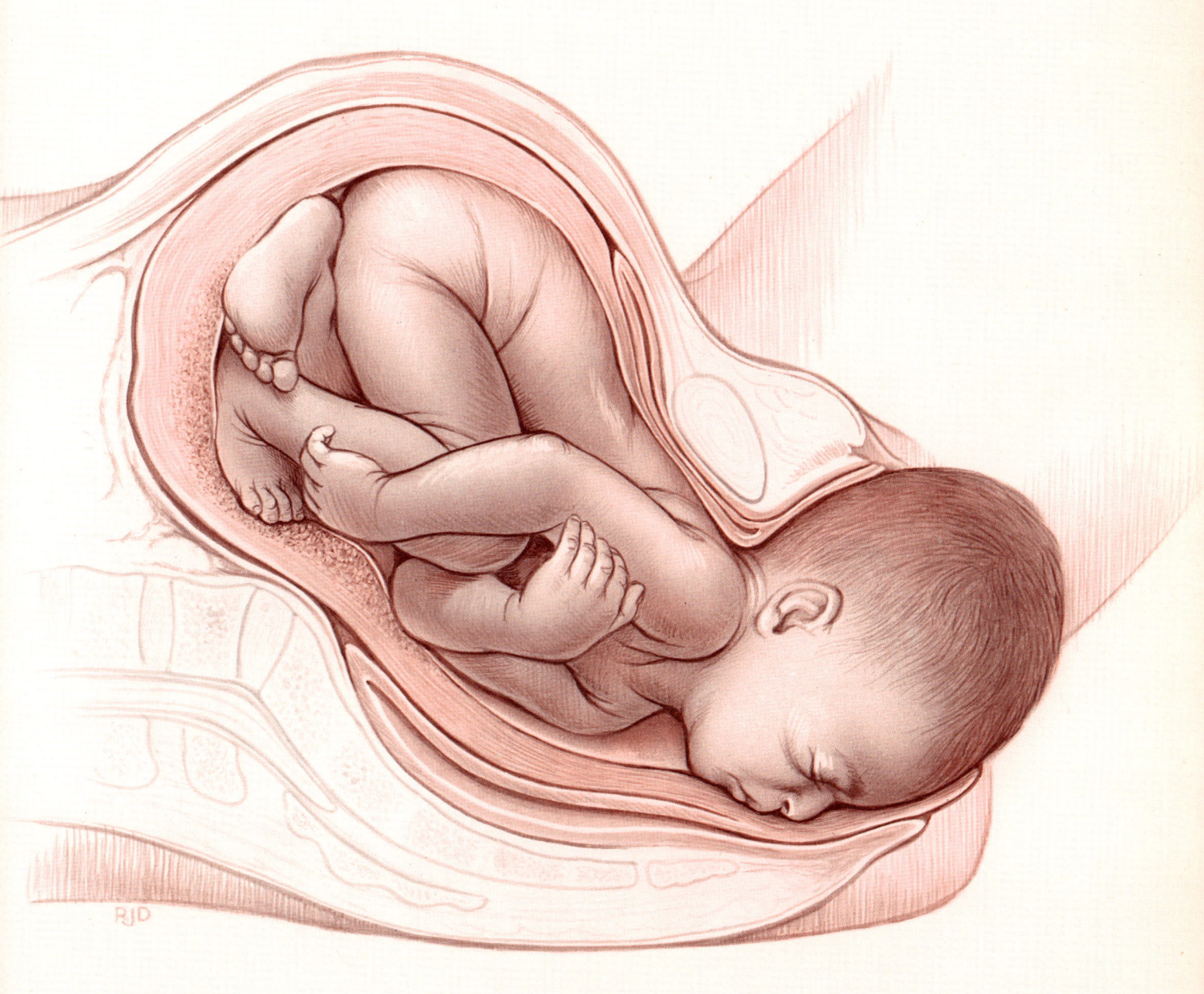

Illustration 43 shows the doctor or midwife assisting delivery of the foetal head. The head has now rotated back toward the position that it was in when the foetus entered the pelvis. Following the delivery of the head, the shoulders, arms, trunk, and legs of the foetus will pass through the birth canal quite easily. Since these parts of the foetus are soft and pliable this part of the delivery is usually accomplished within a minute or two.

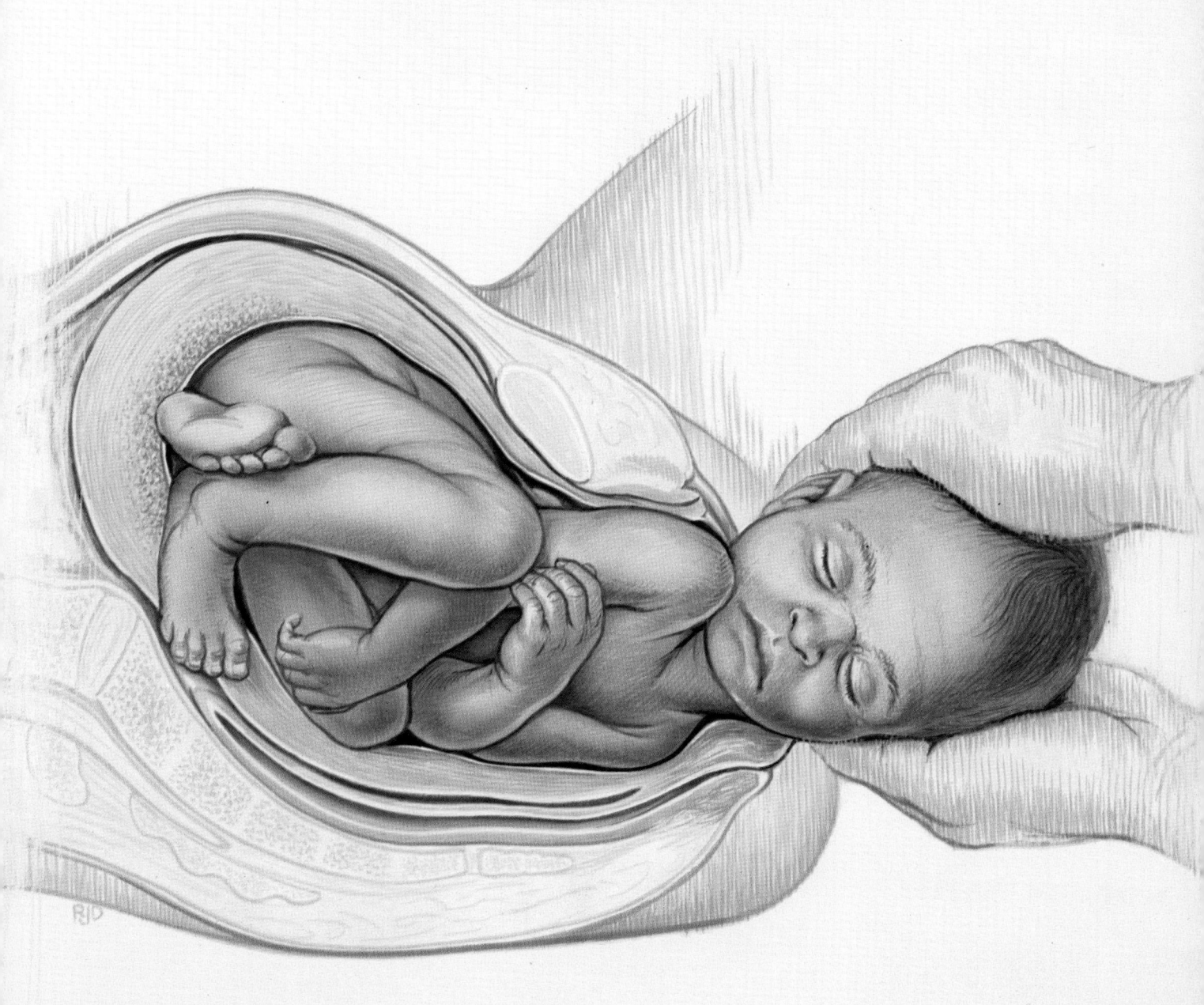

Birth is completed when the baby has been separated from the umbilical cord connecting the newborn infant to the placenta remaining within the uterus. In order to perform this separation, a clamp is attached to the umbilical cord, and the cord is cut. At this point, the baby is usually held by its feet and patted lightly to stimulate it. The infant will usually begin to move, cry, and breathe immediately, and for the first time, the child will sustain life on its own.

44. THE BABY IS DELIVERED

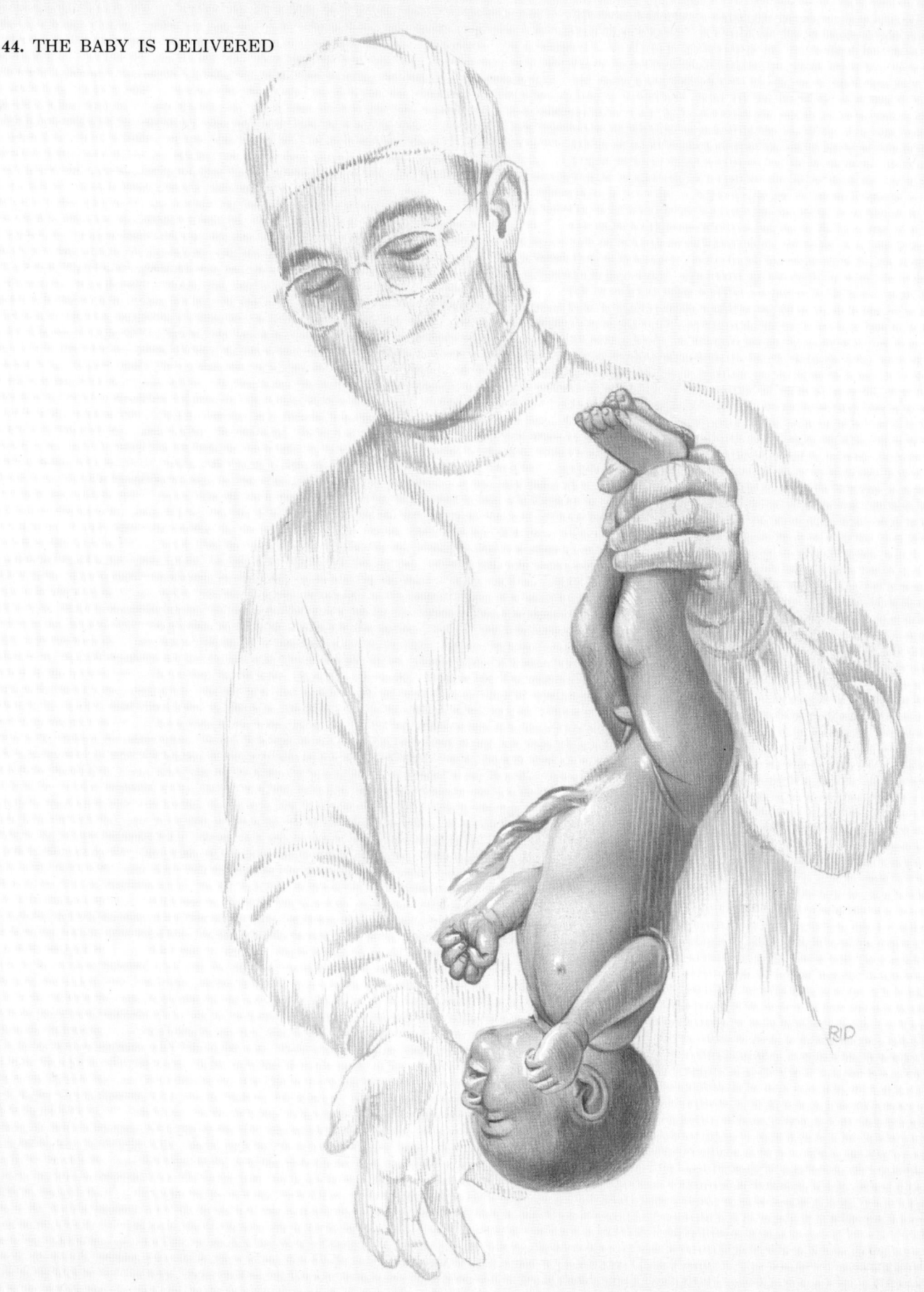

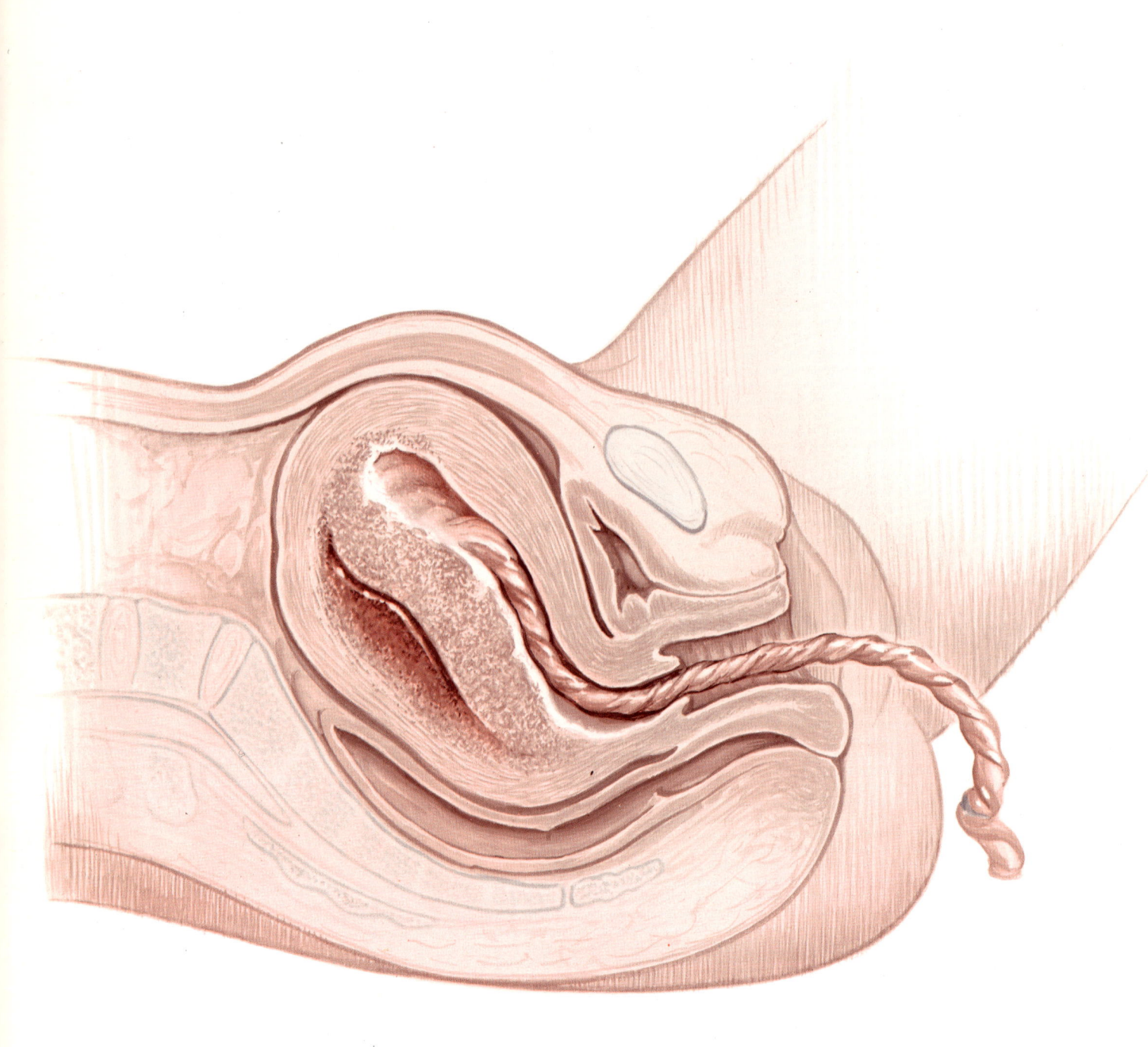

The third stage of labour begins after the baby is born. In this phase the placenta and foetal membranes are expelled, since these structures, called the afterbirth, are no longer of any use and are discarded. This part of the birth process is usually accomplished within ten minutes after the birth of the infant. Often, a small amount of blood accompanies the afterbirth when it is delivered. This blood comes from small blood vessels in the uterine lining where the placenta was attached. These vessels, however, quickly heal and the bleeding usually stops before very much blood has been lost by the mother.

● ● ●

EFFECT OF PREGNANCY AND BIRTH ON THE BREASTS

During pregnancy, the mother's breasts have been changing in order to provide milk for the infant. The changes continue for about three days after delivery, by which time the breasts have become quite enlarged and engorged.

Illustration 46 shows a mother's breast at the time when it is ready to provide milk for the baby. The glands are full and enlarged, and the ducts are ready to release the milk. The issue of milk from the breast is termed lactation.

During pregnancy, the placenta produces hormones which in some way make the breasts ready to provide milk. These hormones, however, do not *cause* lactation. Lactation is started by a hormone produced by the pituitary gland, that is not secreted as long as the placenta is in the mother's body. A hormone produced by the placenta blocks the secretion of this pituitary hormone, until the placenta has been expelled, after which the pituitary gland releases its hormones and lactation begins.

46. LACTATING BREAST

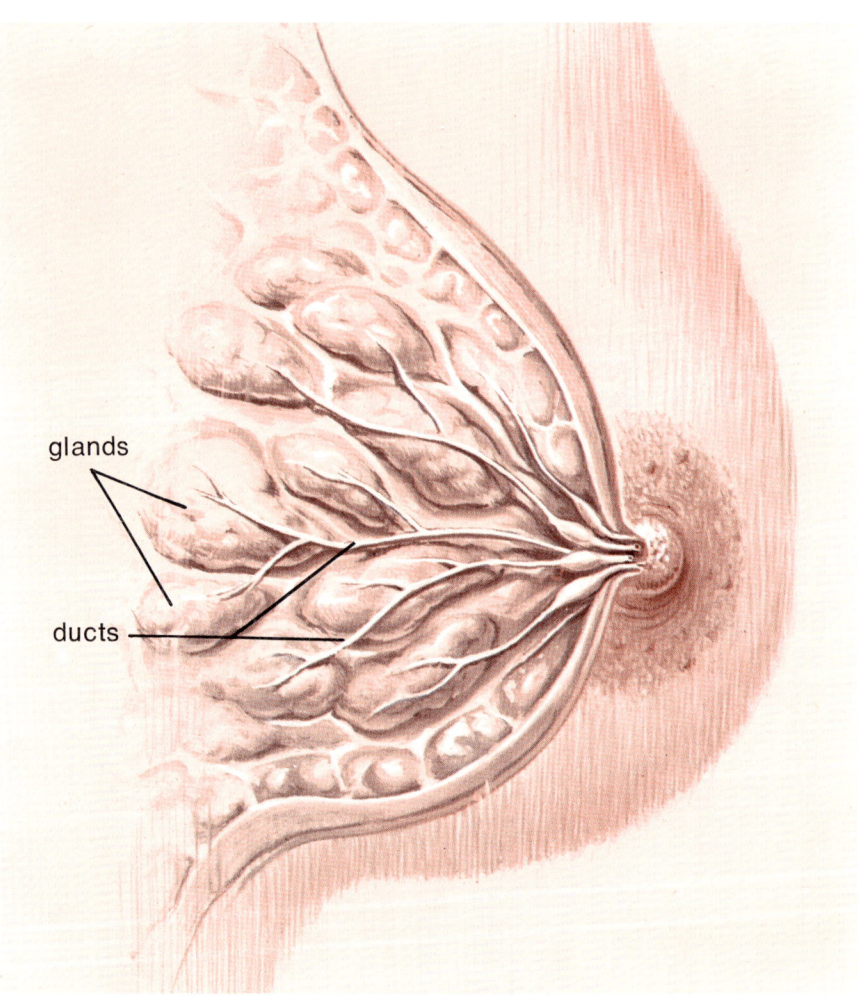

glands

ducts

The production of milk in the breasts occurs in every mother after delivery of the baby, whether she nurses the infant or not. The sucking action of the baby at the breast causes the release of still another hormone by the mother's pituitary gland and this "milk-releasing" hormone acts to force the milk into the duct system of the breasts where it is easy for the baby to get it by sucking.

If the mother nurses the baby, and if she continues to nurse it beyond the postpartum period (six to eight weeks), menstruation may be delayed and so may ovulation and the return of the normal reproductive cycle. Thus the continuation of nursing may sometimes—but not always—block ovulation and provide a degree of contraception during the nursing period.

If the mother does not nurse her baby, milk production will stop within a few days. Sometimes, the doctor will administer a hormone that will produce the same results more rapidly. In either case, the breasts of a nonnursing mother will return to normal in about a month and menstruation will usually begin within six to eight weeks following the delivery. At about this time ovulation also returns to normal.

Following the child's birth, the mother's uterus, cervix, and vagina (which had been stretched during the pregnancy and delivery) soon return to their normal size and resume their normal shape and position within the pelvic area. Several weeks are necessary, however, for the mother's body to completely return to its prepregnant condition. This period following birth is called the postpartum period.

● ● ●

CONTRACEPTION

METHODS: NATURAL AND ARTIFICIAL

Anything that interfers with the fertilization of the egg by sperm can be termed contraception. Anyone who engages in sexual intercourse, but deliberately avoids conception, is employing contraceptive measures. In most general terms, there are six ways that conception can be avoided:

1. By limiting sexual intercourse to a time in the female reproductive cycle when the egg is not available for fertilization
2. By preventing sperm from being deposited in the vaginal canal
3. By blocking the sperm at the cervical mucus and thus preventing them from entering the upper female reproductive tract
4. By keeping sperm from being released into male semen
5. By blocking ovulation, or by preventing the egg from entering the midportion of the tube where fertilization takes place
6. By interfering with the transport of the egg in the tube, or by hindering the process of implantation

There is no one "best" method to prevent conception. The effectiveness of any contraceptive measure depends not only on its degree of protection against pregnancy but also on how well the measure suits the couple utilizing it. Contraception involves social and emotional as well as biological factors. The advice of an interested doctor should be sought to help the couple select the contraceptive method that will be most effective for *them*.

On the following pages, the most commonly used contraceptive methods are discussed and illustrated. In each case, the emphasis is on the manner in which the method interferes with the process of conception—that is, how it acts to prevent pregnancy. Social and emotional factors vary from person to person, and it is not possible to consider these factors here. Furthermore, personal instructions and advice on the many individual features of each method or contraceptive preparation must be obtained from a doctor.

The contraceptive methods discussed here will be grouped into three main categories for convenience of presentation:

1. Methods which interfere with sperm deposition or keep the sperm from entering the cervical mucus
   Natural methods: Withdrawal
                       Rhythm
   Artificial methods: The vaginal douche
                       Spermicidal vaginal
                       preparations
                       The diaphragm or cervical cap
                       The condom
2. Methods which inhibit ovulation
   Oral contraceptives: Combination pills
                       Sequential pills
3. Method where mode of action is still unknown
   Intrauterine devices

Not all methods are equally effective. A list comparing the relative effectiveness of each method appears on page 111.

METHODS WHICH INTERFERE WITH SPERM DEPOSITION OR KEEP THE SPERM FROM ENTERING THE CERVICAL MUCUS

Withdrawal, which in medical terms is called coitus interruptus, is one of the simplest and oldest natural methods of preventing pregnancy. With this method, the male withdraws the penis from the vagina just before ejaculation, thus preventing semen from being deposited within the vaginal canal. The method is effective only if the male is able to completely avoid ejaculation within the vagina, for if even the smallest amount of semen containing sperm is deposited in the vagina, fertilization could conceivably take place.

The *rhythm method* is the second natural method of contraception. With this, couples do not engage in intercourse during the period of the female reproductive cycle when the fertilizable egg is present in the fallopian tube. Since the time when the egg is available for fertilization is a relatively short period in the reproductive cycle, intercourse during any other time is considered safe and pregnancy is not likely to occur.

In theory, the rhythm method is quite simple. For several reasons, however, it is not a very reliable form of contraception. The chief reason is that sperm can survive within the cervical mucus for as long as forty-eight hours, hence, intercourse even two days prior to ovulation might possibly result in conception. Also, after ovulation, the egg is available for fertilization for approximately twelve to twenty-four hours, thus extending the fertile period for an additional day. In the case of a female with a *perfectly regular* twenty-eight-day cycle, the minimum period for avoiding intercourse (in order to avoid pregnancy) might be as short as from $2\frac{1}{2}$ days before ovulation to $1\frac{1}{2}$ days after ovulation. Illustration 48 shows such a "safe" period and such a "danger" period when using the rhythm method, based on a twenty-eight-day reproductive cycle as described above.

48. RHYTHM SYSTEM FOR A 28-DAY CYCLE

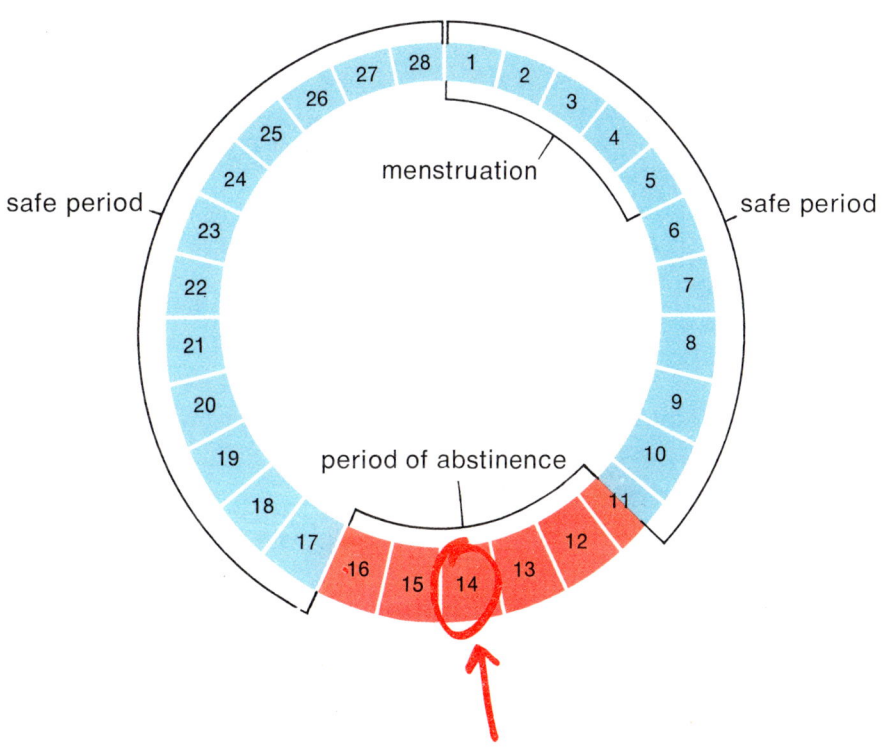

Another unfavourable factor in the rhythm method is the difficulty of determining the exact time of ovulation. If the length of the menstrual cycle varies slightly from month to month, the exact time of ovulation will also vary. For example, if the menstrual cycle is twenty-eight days, ovulation will occur on the fourteenth day, but if the cycle is only twenty-six days ovulation will occur two days earlier, or on the twelfth day. If the cycle is longer—say, thirty days—ovulation will occur on the sixteenth day.

Since the length of the menstrual cycle and the time of ovulation are not entirely predictable, the rhythm method based on the calendar alone (called the calendar rhythm method) should be used with great caution. Most doctors who supervise women using this method of contraception suggest that they keep careful records of their menstrual cycle for at least one year, and then figure the "safe" period based on the longest and shortest cycles during that year. For example, if the menstrual cycles ranged from twenty-six to thirty days during the previous twelve months, one could expect ovulation to occur at any time between the twelfth and the sixteenth days of the present cycle. The woman would then avoid intercourse from $2\frac{1}{2}$ days before the twelfth day until $1\frac{1}{2}$ days beyond the sixteenth day. Tables and charts have been developed to determine safe periods within normal ranges.

It should be stressed that for maximal effectiveness the rhythm method should be supervised by an interested doctor.

The *temperature rhythm method* allows the female to predict the time of ovulation more accurately than the calendar method. This system is based on the fact that for most females there is a rise in body temperature just after ovulation.

A special thermometer and a graph, called the basal body temperature (BBT) graph, are used with this method of contraception. The sample BBT graph in Illustration 49 shows the rise in temperature that occurred following ovulation and continued during the remainder of the menstrual cycle. If intercourse is deferred until the temperature rise occurs, there is little danger of pregnancy. The reason for this is that by the time the temperature increase is recorded, the egg is past the point where it can be fertilized. Also, the action of progesterone, which is being secreted by the *corpus luteum* (after ovulation), makes the cervical mucus unfavourable for sperm penetration.

Although the temperature rhythm method will determine the safe period after ovulation, it will not indicate the safe period *before* ovulation. To determine this, the female should record the time of ovulation for several cycles to establish variations in ovulation, just as she would do with the calendar rhythm method. Many doctors feel that for maximal effectiveness features of the calendar rhythm method should be combined with those of the temperature rhythm method.

49. BASAL BODY TEMPERATURE GRAPH

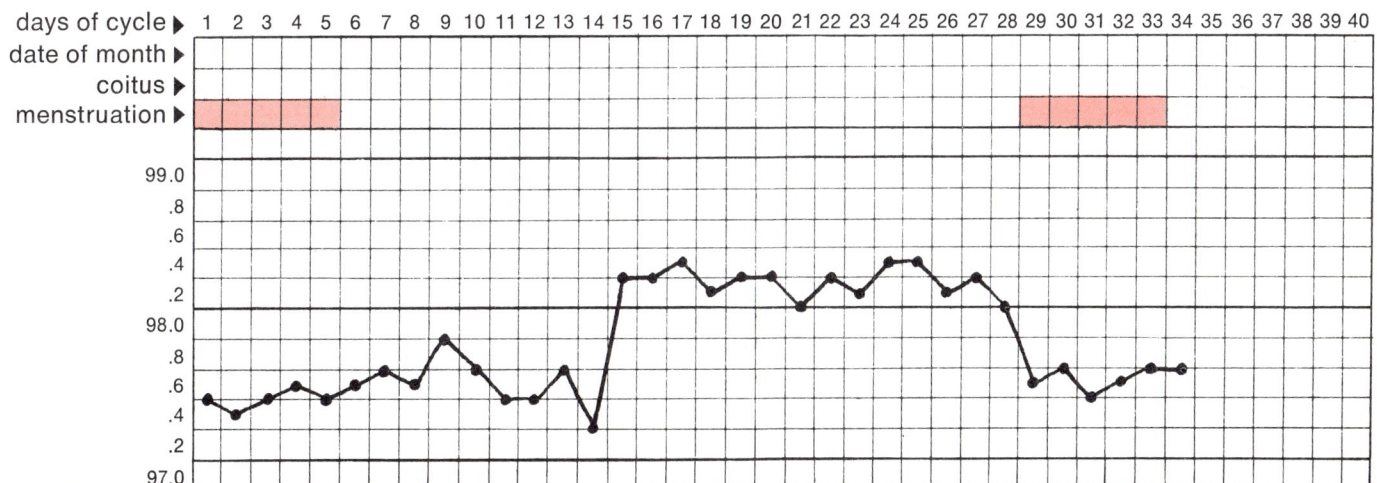

In the category of artificial methods, the *vaginal douche* is perhaps one of the oldest but also one of the least effective. With this method, a relatively large amount of tap water (1 or 2 quarts) is sprayed into the vagina with a fountain or bulb syringe immediately following intercourse. The theory is that flushing the vaginal canal will wash away the vaginal pool of semen, but since active sperm may reach the cervical mucus within a minute or two after unprotected intercourse, the contraceptive effectiveness of douching is very low. In fact, some doctors consider this method so ineffective that it should not be seriously considered as a contraceptive technique.

Spermicidal vaginal preparations contain a chemical that will kill the sperm or make them incapable of penetrating the cervical mucus. These preparations are available as foams, creams, jellies, or tablets. Most of them come in a tube or aerosol dispenser, with a special applicator for depositing the material in the upper vaginal canal prior to intercourse. Illustration 50 shows a vaginal applicator filled with a spermicidal product which is being deposited in the vagina. Vaginal foam tablets are inserted without an applicator and dissolve in

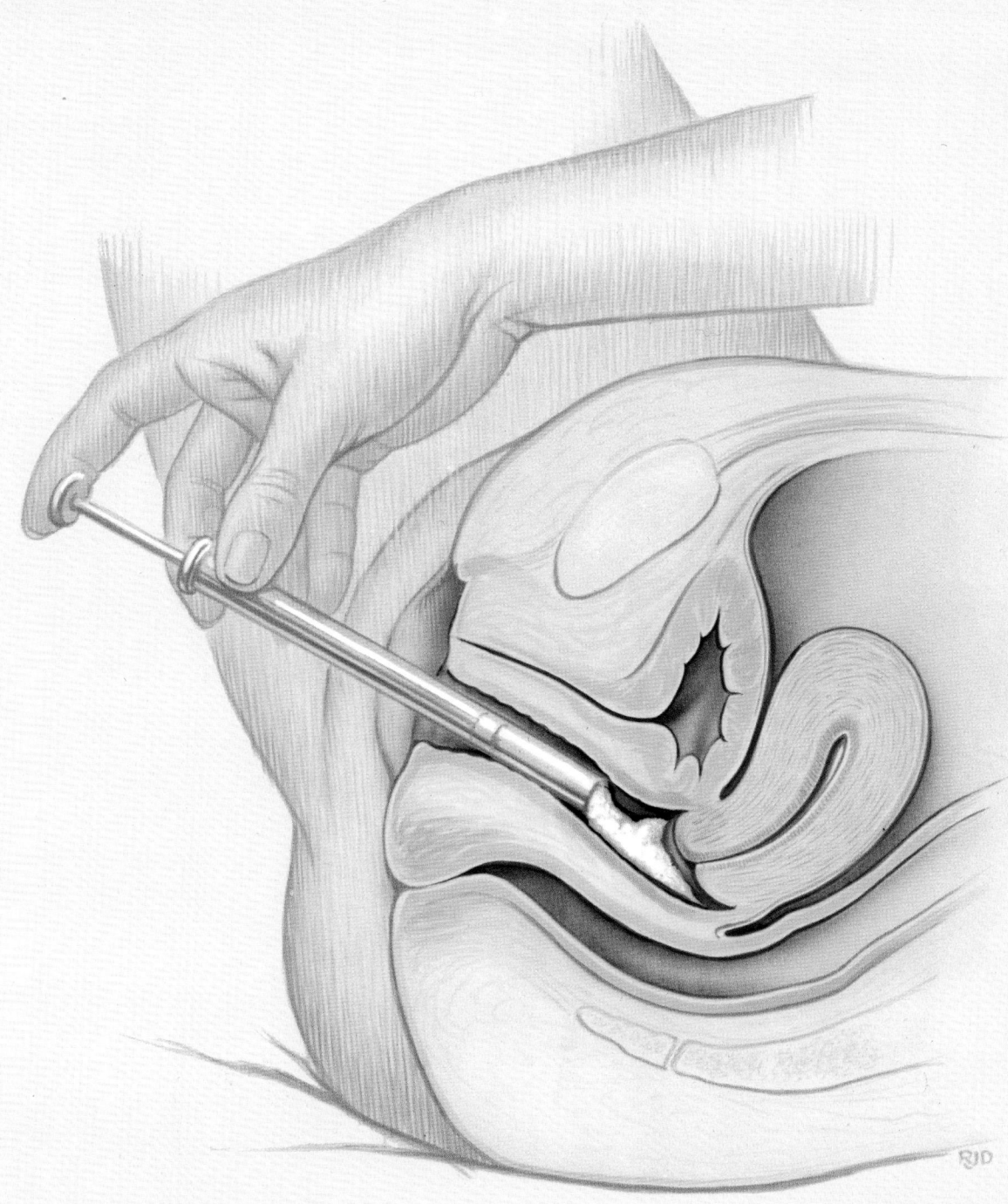

normal vaginal secretions producing the foam barrier. Illustration 51 shows the upper vaginal canal and cervix immediately after intercourse, where a contraceptive foam preparation has been used. Note that the chemical has been spread over the cervix. In addition to killing sperm, the spermicidal preparation covering the cervix also reduces the possibility of sperm penetrating the cervical mucus.

Spermicidal suppositories melt at body temperature to release their chemical agents. The suppositories are similar in appearance to the foam tablets.

The effectiveness of these preparations in preventing pregnancy is moderately high if they are used regularly and if there is a fresh application approximately five to ten minutes before each coital act. The chemicals in them may occasionally cause an irritation of the vagina but this is rarely serious. The foams, jellies, and creams also serve as a mild lubricant during intercourse.

51. SPERMICIDAL FOAM BARRIER IN VAGINA

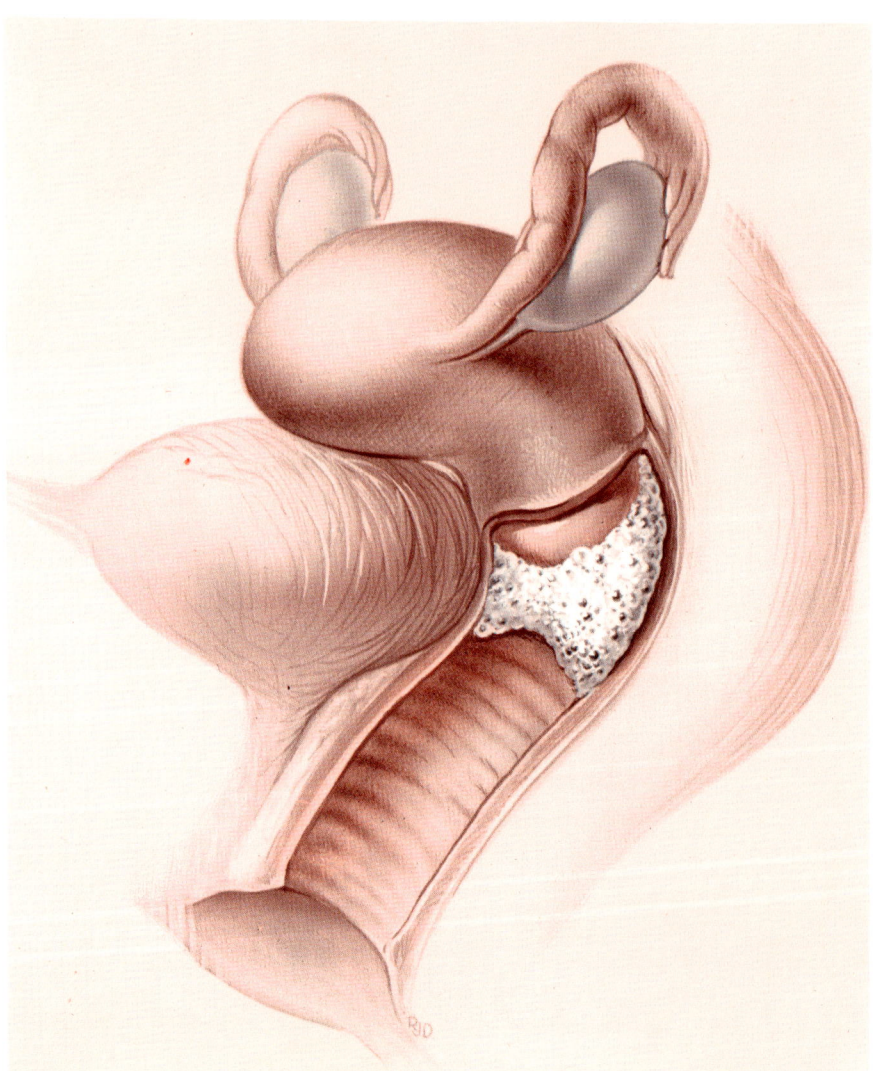

The *diaphragm and cervical cap* are examples of mechanical devices which cover the cervix of the female to prevent sperm from entering the cervical mucus and the upper female reproductive tract. These devices range from a simple vaginal sponge placed against the cervix to a tight-fitting plastic cap which fits over the cervix.

A highly effective and widely used mechanical method of female contraception is the rubber vaginal diaphragm. Illustration 52 shows a typical soft rubber diaphragm in its protective case. Illustrations 53 to 55 show how the diaphragm is inserted and how the shallow cup is positioned over the cervix.

Diaphragms come in various sizes according to their diameter in millimeters, the average being 75 mm. A doctor determines the size of diaphragm the individual needs and provides instructions for its use. For most persons, it is recommended that it be used with a spermicidal cream or jelly placed around the rim and in the shallow cup before positioning against the cervix.

52. RUBBER DIAPHRAGM

The rim of the diaphragm is made of a flexible metal spring which permits it to snap back into shape after it has been folded for insertion into the vagina. Illustration 54 shows the folded diaphragm being inserted into the vagina with the fingers (plastic inserters are also available). In Illustration 55 the rubber diaphragm is shown *in position* in the vaginal canal. Since the vagina will stretch, the diaphragm will take its normal circular shape and will cover the cervix with a thin layer of rubber. If a spermicidal preparation is used with the diaphragm, it will spread over the cervix, adding a second—chemical—barrier to prevent sperm from reaching the cervical mucus.

53. DIAPHRAGM FOLDED FOR INSERTION

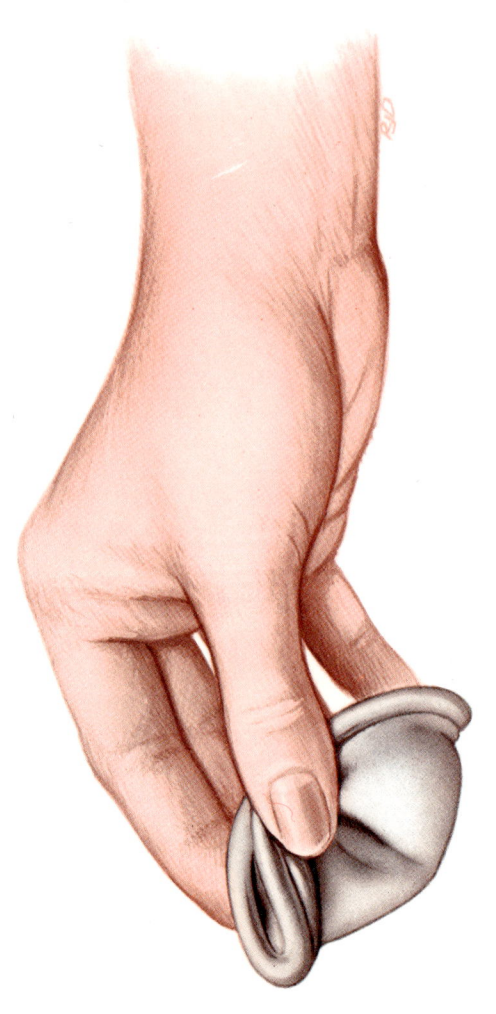

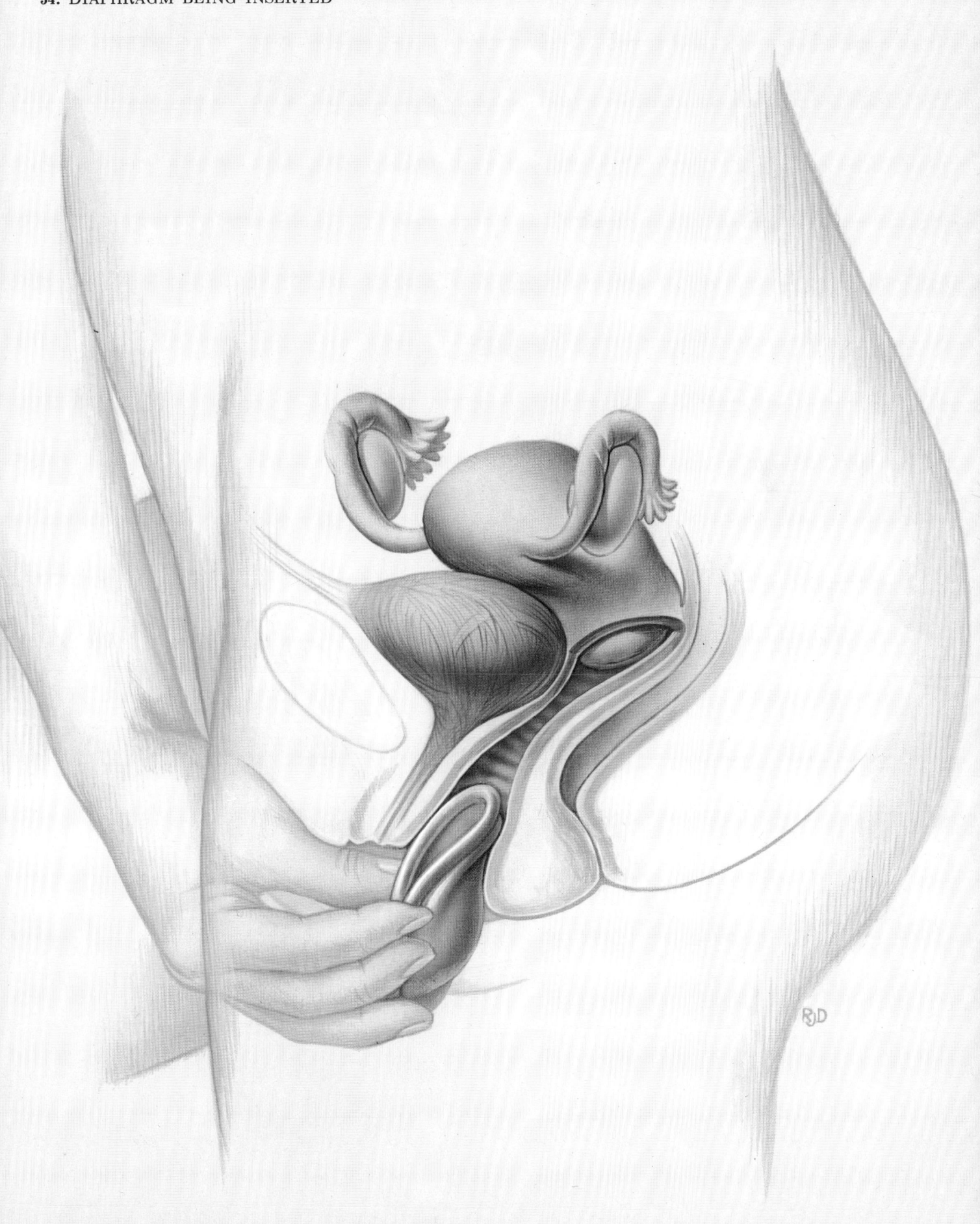

When the diaphragm is in place, it is held there by the tension of the metal rim. The bottom edge of the rim rests in the posterior vagina and the upper edge against the vaginal wall behind the bladder and the pubic symphysis.

The diaphragm may be put into place several hours before intercourse and may be left in place for as long as twenty-four hours. When it is in place, the wearer does not feel it and she can withdraw it easily, allowing it to collapse as it is removed from the vaginal canal. It is then washed, dried, and returned to its protective case for reuse.

A widely used male contraceptive is the *condom*, which acts to prevent the deposition of semen in the vagina by enclosing the erect penis in a rubber or animal-skin sheath. The condom measures 7 or 8 inches in length and fits snugly over the erect penis. Its end is sealed so that on ejaculation the semen is deposited without entering the vagina. Most condoms used today are made of animal intestine and may be washed and reused.

The condom is extremely effective if it does not rupture during intercourse. If used along with a lubricating spermicidal preparation, the degree of contraceptive protection is even higher.

● ● ●

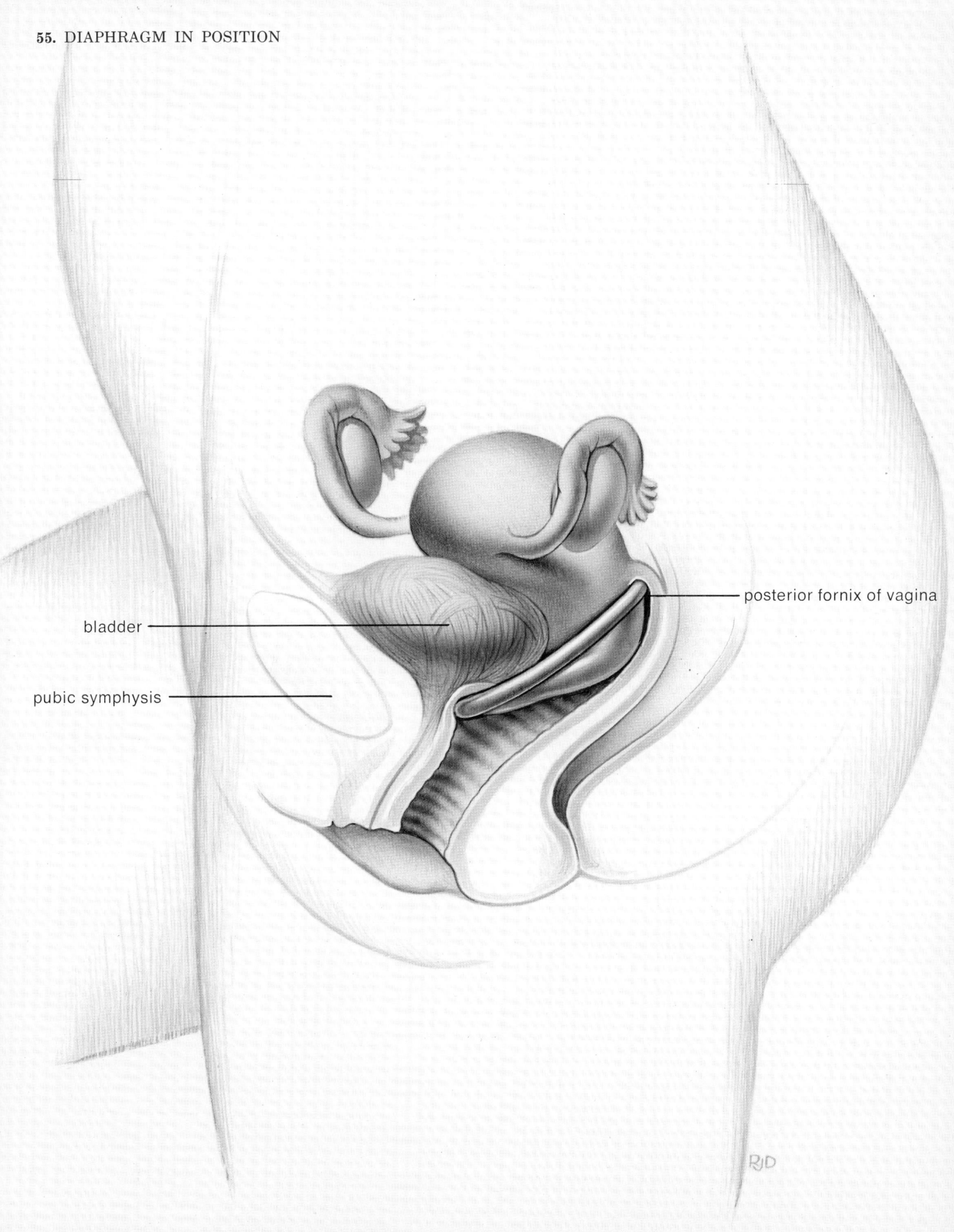

posterior fornix of vagina

bladder

pubic symphysis

METHODS WHICH INHIBIT OVULATION

Oral contraceptives, unlike those previously described, prevent fertilization by inhibiting the release of the egg from the female ovary and/or by making the wall of the uterus unreceptive to egg implantation. Thus with this method, sperm can be allowed to be freely deposited in the vagina and enter the female reproductive system.

As stated earlier, there are certain times in a woman's life (during pregnancy and nursing) when natural hormones block ovulation. Oral contraceptives contain artificial (synthetic) hormones much like natural hormones which block the output of the pituitary gonadotrophic hormone that normally causes ovulation (See page 34). In addition, the synthetic hormones in oral contraceptives stimulate the uterus so that a regular menstrual cycle is achieved. This dual action is shown in Illustration 56.

Two main types of oral contraceptive pills are available, and they work in the same way. First are the *combination pills*, which contain two synthetic hormones (oestrogen and progesterone) in one tablet. The second type are the *sequential pills*, which are packaged so that the woman takes tablets containing a synthetic oestrogen hormone for the first three-quarters of her medication cycle and tablets containing synthetic oestrogen and progesterone for the remainder of the cycle.

With the sequential type of oral contraceptives, while suppressing ovulation an attempt is made to parallel normal changes in the cervical mucus and endometrium during the menstrual cycle. With the combination pill, changes take place in which the endometrium and cervical mucus appear only as they would during the postovulatory phase of the menstrual cycle. In addition, the cervical mucus and the endometrium do not favour either sperm transport or egg implantation.

While both types of oral contraceptives are highly effective, there seems to be a slightly greater chance of pregnancy in women using the sequential pill. If the pills are taken as directed, however, the chance of pregnancy is very low. For example, doctors have estimated that for a group of one hundred women using oral contraception for a one year period only about one pregnancy might occur.

Most oral contraceptive pills are packaged by the manufacturer so that twenty or twenty-one pills are taken on a one-a-day schedule beginning on the fifth day of the menstrual cycle. Most doctors now prescribe the twenty-one-day regimen. Menstruation will usually begin three days after the last pill is taken. Then, on the fifth day of the menstrual cycle, the next series of pills is begun. Some companies package their pills so that the woman takes one tablet every day during the complete twenty-eight-day cycle. With such a package, the last seven or eight pills usually contain no medication, so the effects of the synthetic hormones in the pills containing medication are the same as with the interrupted cycle.

With all oral contraceptive pills, menstruation is usually perfectly regular, occuring every twenty-eight days. Should the woman stop taking oral contraceptives, ovulation will usually begin two to six weeks later and her menstrual cycle will then return to its premedication pattern.

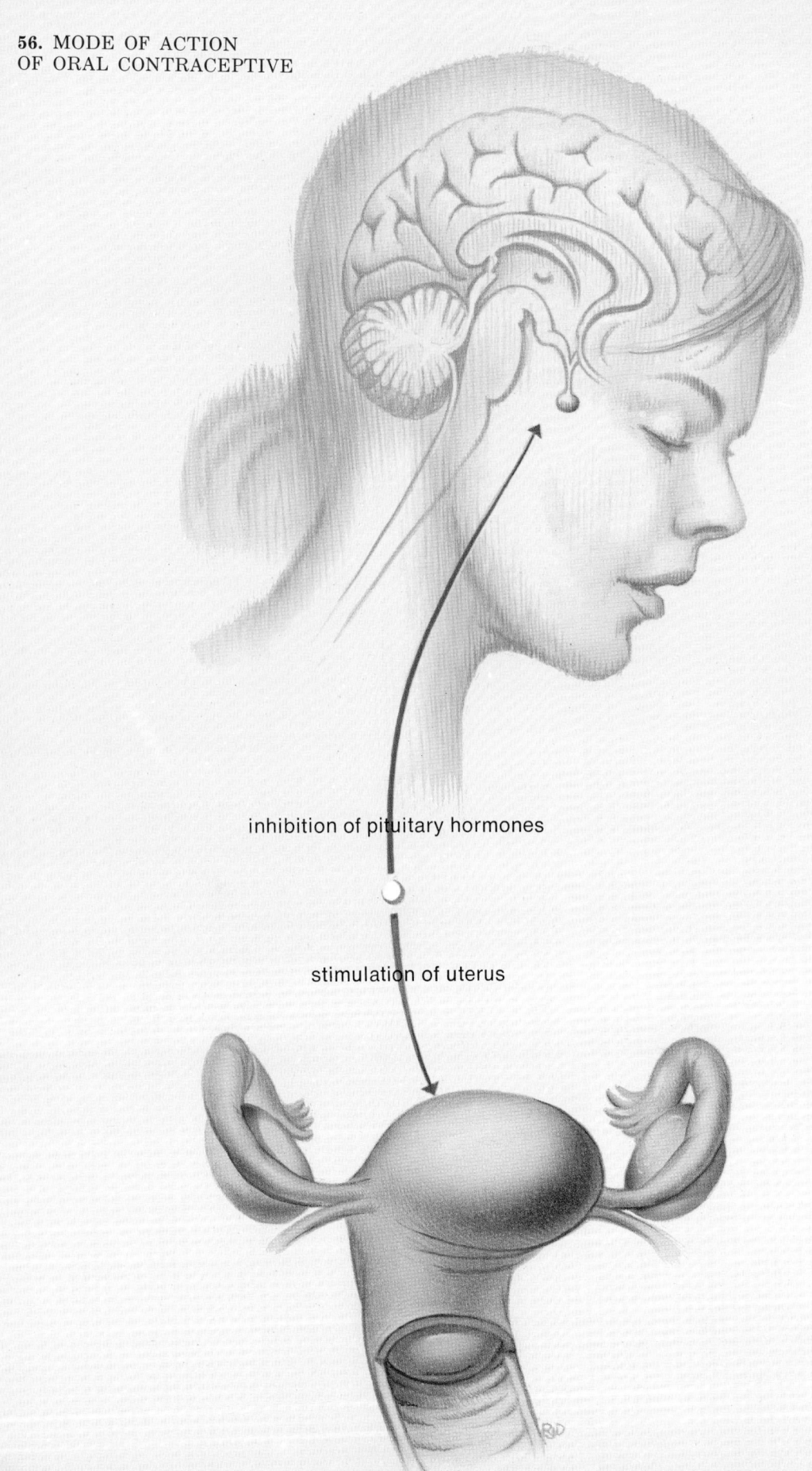

inhibition of pituitary hormones

stimulation of uterus

Since oral contraceptives contain potent hormonal preparations, they must be prescribed by a doctor. A woman taking these contraceptives should visit her physician regularly because the drugs in the pills may cause disorders or side effects that require medical attention. Such side effects as stomach and bowel distress, irregular vaginal bleeding, and an increase in weight have been reported in women using both the combination and the sequential pills. There is also some evidence that oral contraceptives may cause disorders relating to blood coagulation and circulation, and therefore a careful medical history is usually taken by the doctor before the pills are prescribed. Oral contraceptives have been available for only about ten years and, as with any other medication, it takes many years before the long-term safety of the product is firmly established.

It should be noted for comparison that as many as 30 per cent of women who initially select oral contraception may also abandon the method or change to another method after the first year of use.

● ● ●

METHOD WHERE MODE OF ACTION IS STILL UNKNOWN

Most methods of contraception rely on keeping the active sperm and the egg separate during the period when fertilization can take place. Doctors have known for many years, however, that the presence of a plastic or metal object in the uterus would prevent pregnancy. Just how, or why, such an object works, no one is sure.

Some doctors suggest that such intrauterine devices cause the egg to pass down the fallopian tube too rapidly to be fertilized, or if it is fertilized, it arrives in the uterus at a stage in development when implantation is not possible. It may be that the device interferes with implantation of a fertilized blastocyst, or even affects the cyclic growth of the uterine endometrium. Such devices, however, have no marked effect on the normal menstrual cycle.

57. INTRAUTERINE DEVICES (SLIGHTLY REDUCED)

In recent years, a variety of metal and plastic intrauterine devices (IUDs) or intrauterine contraceptive devices (IUCDs) have been developed and tested and have proved to be effective. Illustration 57 shows that IUDs may be loops, rings, spirals, bows, or a combination of these forms. They are shaped in these ways so that they will not fall out of the uterus or be expelled from it once they have been inserted.

A doctor inserts the IUD by straightening it in a tubelike instrument. The instrument is passed through the cervix of the uterus, and the IUD is pushed out the end of the tube. When the IUD is free in the uterine cavity it assumes its original shape. Occasionally uterine contractions will cause the IUD to be expelled, and sometimes a doctor must remove one because it causes discomfort or bleeding. If it is not removed or expelled, it can remain in place for many months, or even years, with no harm or discomfort to the woman, and can be removed by a doctor at any time.

58. INSERTION OF INTRAUTERINE DEVICE

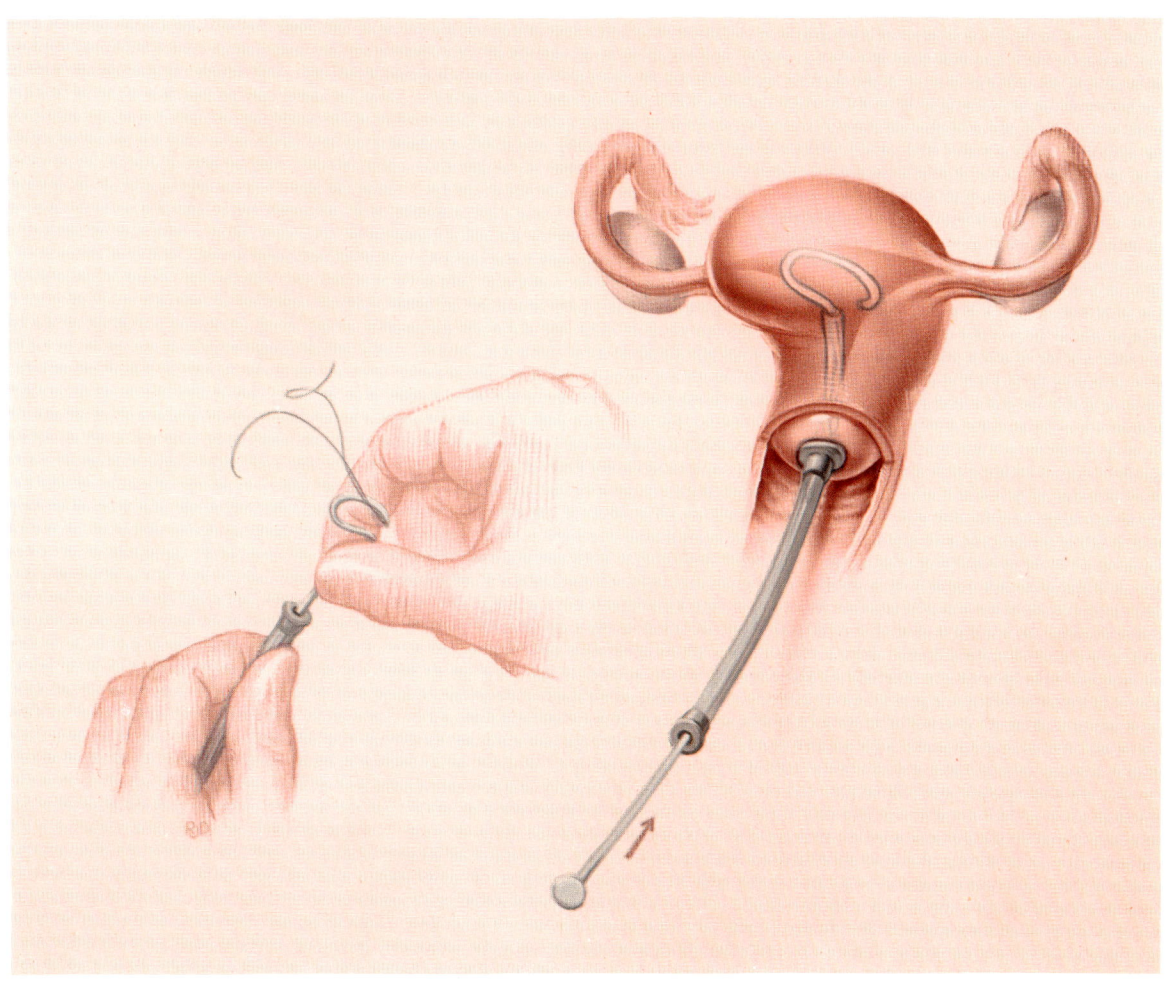

The cross-section view of the uterus in Illustration 59 shows the IUD in place. Some intrauterine devices, such as the one shown here, have a piece of plastic attached to them which extends down through the cervix. This helps to indicate the presence of the device and its type, and makes it easier for the doctor to remove it.

As long as the intrauterine device is in place, it is an effective contraceptive measure. Only about 80 per cent of the women who have had IUDs inserted, however, have been able to use this method of contraception on a long-term basis. The remaining 20 per cent either expelled the device or asked to have it removed because of discomfort or bleeding. Some doctors feel that young women who have never had a child are more likely to have difficulty with IUDs than those who have given birth, perhaps because of the smaller size of the uterus which has never been stretched by child bearing. Women in their twenties can frequently wear an IUD quite successfully regardless of childbearing status.

59. INTRAUTERINE DEVICE IN PLACE (ACTUAL SIZE)

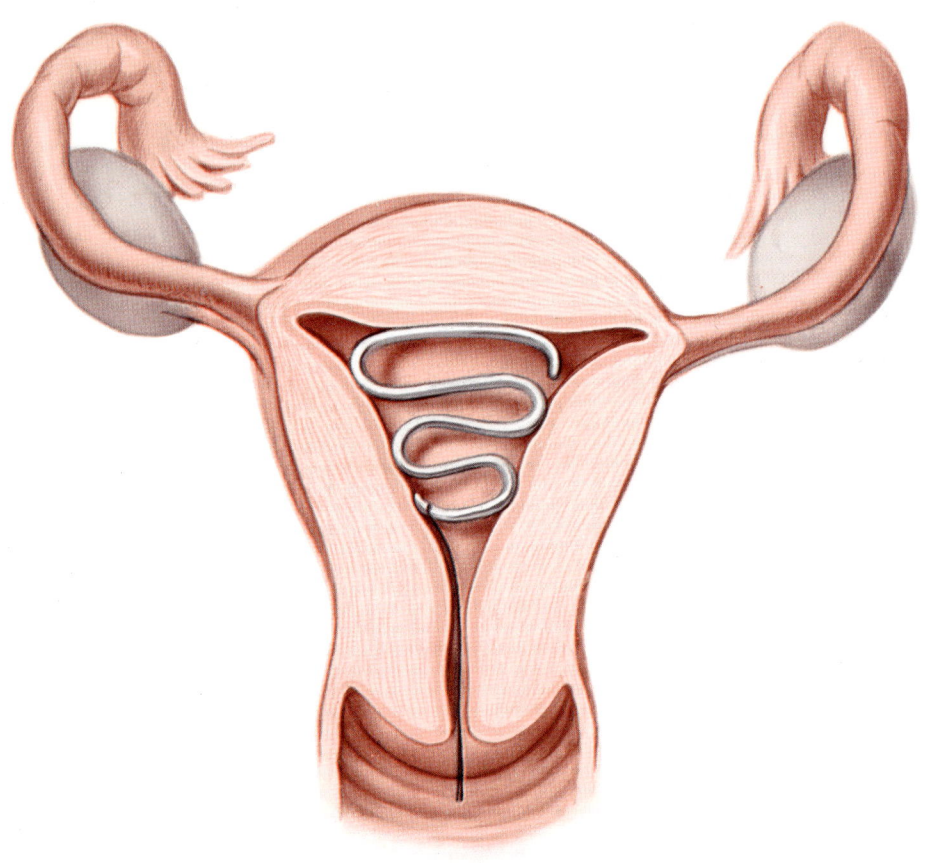

In the list below, the commonly used contraceptive methods are grouped in order of their relative effectiveness. Within the groups the methods are simply listed alphabetically.

Most effective:	Oral contraceptives (combination or sequential type)
Highly effective:	Condom
	Diaphragm with spermicidal jelly or cream
	Intrauterine devices
Moderately effective:	Rhythm (temperature method)
	Spermicidal vaginal preparations
Less effective:	Rhythm (temperature method)
	Spermicidal vaginal preparations
Less effective:	Rhythm (calendar method)
	Withdrawal
Least effective:	Vaginal douche

EXPERIMENTAL METHODS OF FERTILITY CONTROL

Despite the many contraceptive methods available, no one method is ideal for all individuals, nor is there any one method that can be considered perfect. For these reasons, doctors continue to search for other methods to prevent pregnancy.

Currently being tested are oral chemical agents that affect the cervical mucus and block sperm penetration, thus keeping the sperm out of the upper reproductive tract of the female where fertilization takes place. These drugs contain a very low dose of a progesterone-like hormone which does not affect ovulation. Like oral contraceptives now available, they are usually taken on a twenty-day cycle, a twenty-one-day cycle, or throughout the entire menstrual period.

Also being studied are injectable hormones that can be given to the female just once a month, or once every three months (some are being tested that require only one injection a year). These preparations slowly release hormonal medication that either stops ovulation or blocks sperm penetration by changing the consistency of the cervical mucus.

In some studies hormones embedded in a type of plastic (silastic) are being implanted in the woman's skin. The tablet slowly releases hormones—for months or even years—and acts much like the injection technique mentioned above. The implant can easily be removed at any time, should the women desire to become pregnant. It may also be possible simply to place a hormone-impregnated plastic device within the vagina and leave it in place for months at a time.

Doctors are also studying the possibility of providing mini pills, injectable hormones, or hormone tablets for implantation for use by men. These preparations would work much like those used by women, but would inhibit the production of sperm or make sperm immobile or incapable of fertilization.

Many men and women have said that they would like a pill that could be taken *after* intercourse to prevent pregnancy. Such pills are now being studied. They are hormone-type preparations which act to interfere with tubal transport of the egg and/or implantation.

Another approach to contraception that is in the very early testing stage is the immunological method. With this type of contraception, the person would receive something like a vaccination which would either inhibit spermatogenesis in the male or would block ovulation or sperm transport and fertilization in the female.

What the future holds in the way of contraceptive measures cannot possibly be known at this time. Many factors will exert an influence, and so many things are still unknown in the areas of medicine, psychology, sociology, and economics that suggestions for future approaches would be nothing more than guesswork.

One far-in-the-future idea is a medication that all men or women would take continually throughout their reproductive years—perhaps have it in one of their basic foods, as iodine is now in salt. Then, when fertility is desired, the person would simply take an antidote. It may never be possible to develop such a product, but the idea serves to emphasize that future methods of contraception will depend on the needs of the people and the ingenuity and imagination of the doctors who work in this area of medicine. There is no reason to believe that future methods of contraception will simply be an extension or perfection of present methods.

STERILIZATION

Contraception, as it has been discussed in the preceding pages, is merely a method of temporarily inhibiting conception. It does not affect the person's ability to reproduce.

Sterilization, however, is a procedure which renders the person incapable of reproduction. For the most part, it is a surgical technique which, once accomplished, is difficult (and often impossible) to reverse. Sterilization, as a contraceptive measure, is never undertaken until the person involved has had the number of children he or she desires.

Male sterilization (called a vasectomy or vas ligation) is a surgical procedure which interrupts the pathway of the sperm and is accomplished by either cutting or tying the *vas deferens*. Illustration 60 shows how this is done. After the doctor has injected a pain killer into the surrounding tissue, he makes a small cut in each scrotal sac. The *vas deferens* on each side is drawn out through the incision and the ducts are cut. This procedure blocks the sperm so that they can no longer pass along the *vas deferens* and into the urethra in the normal fashion (see page 9).

Following this operation the male cannot release sperm, but the other secretions of the male reproductive tract are unaffected. During intercourse he will have no difficulty either achieving an erection or ejaculating semen; the semen, however, will contain no sperm.

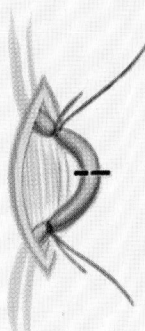

a. vas being ligated and divided

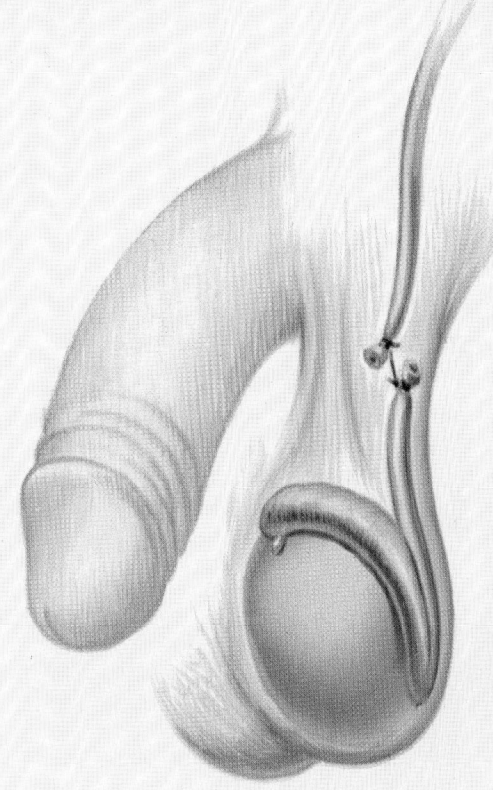

b. final appearance

Since the female bears the physical strain of pregnancy, sterilization of the female may be undertaken for medical reasons if the doctor feels that a pregnancy would seriously jeopardize the woman's health. Or, the woman may request such an operation after she has had the number of children she desires.

The most common operation for female sterilization, called tubal ligation, is more involved than the vasectomy. Often, it is performed while the woman is still in the hospital after giving birth, although it can be done at any time. To perform the tubal ligation, the doctor makes an incision in the woman's abdominal wall, cuts both fallopian tubes, and usually removes the middle section of each tube. Illustration 61 shows the appearance of the tubes after a tubal ligation has been performed. Even though ovulation will continue, it is now impossible for the egg to pass beyond the blocked point in the tube. Also, sperm entering the tube from the uterus are unable to pass beyond the point of the operation. Since sperm and egg cannot meet, fertilization is impossible. It should be noted that on occasion partial or total removal of the woman's uterus has to be performed by her doctor and this also serves as a method of sterilization.

The vasectomy and the tubal ligation are extremely effective contraceptive measures. Once these operations have been performed, however, it is very difficult to reverse the procedure should additional children be desired.

New operations have been developed to avoid opening the abdominal cavity of the female when performing tubal ligation. One such technique involves a small telescope-like instrument that is inserted into the pelvic cavity and is used to both cut and tie off the fallopian tubes.

For the male, studies are being done whereby the *vas deferens* is cut, and a small piece of plastic tubing with a plug in it is attached to both ends of the severed tubes. The plug blocks the passage of sperm. Should the man desire, the plug can later be removed so that the sperm can again pass through the *vas deferens* and out the urethra.

61. TUBAL LIGATION, FINAL APPEARANCE

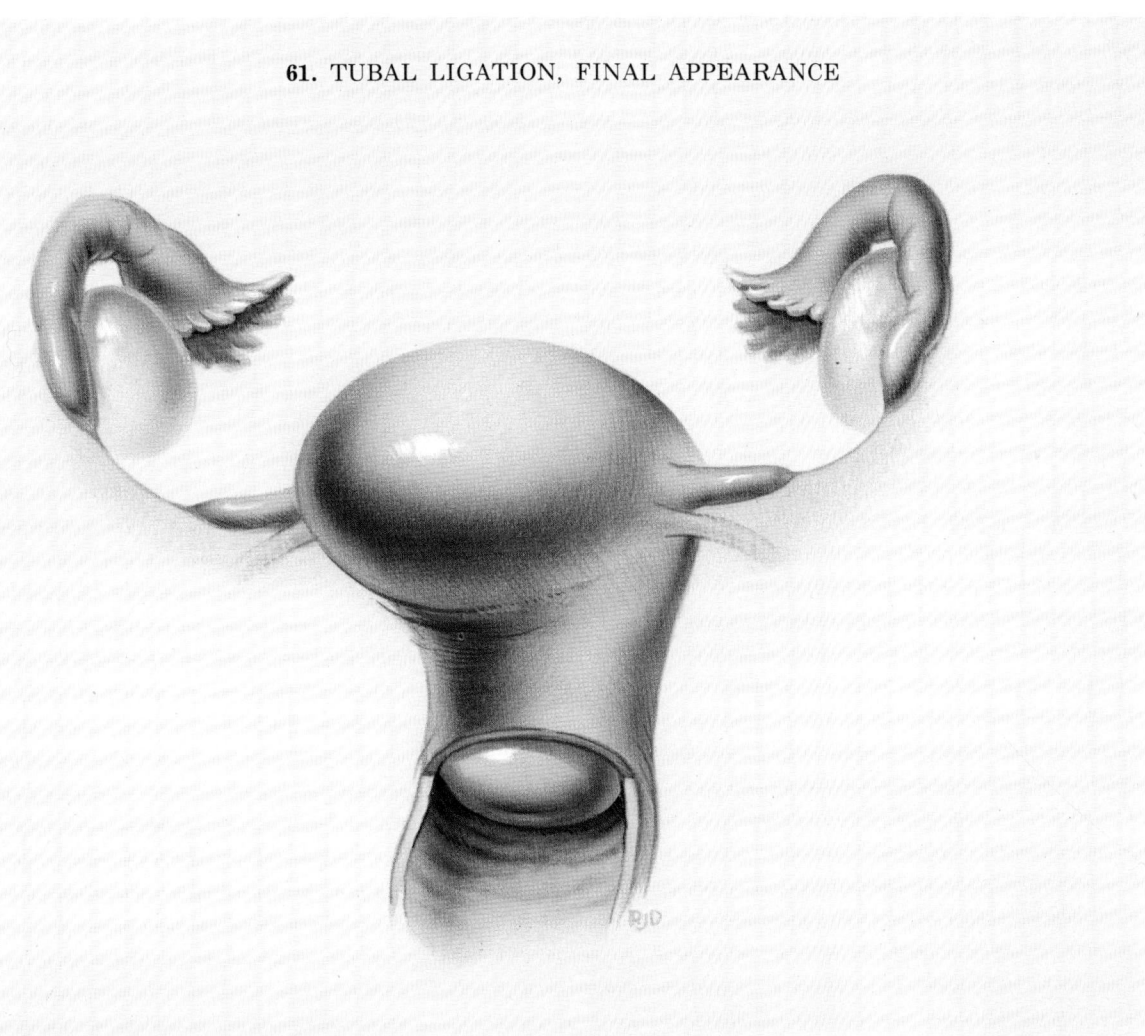

MENOPAUSE

In the female, usually sometime between forty-five and fifty years of age, ovulation ceases and the remaining follicles and eggs in the ovaries degenerate. When this happens, a series of changes occurs within the female reproductive system which signals the end of the reproductive function. This period of change is called the menopause or, sometimes, the "change of life."

Ovulation stops, and the production of the hormone oestrogen is greatly reduced. This has a direct effect on the rest of the female's reproductive system. The menstrual cycle ceases and the woman often experiences "hot flushes," "flashes," and excessive perspiration. The vaginal lining becomes thinner and vaginal secretions decrease. The dryness of the vagina may often make intercourse uncomfortable, but the doctor can give the woman synthetic oestrogen to relieve the symptoms. This will not cause ovulation to begin again, however.

Illustration 62 shows the female reproductive tract after menopause has occurred and indicates the general shrinkage of all the reproductive structures. For comparison, see Illustration 9, which shows the female reproductive organs during the fertile years.

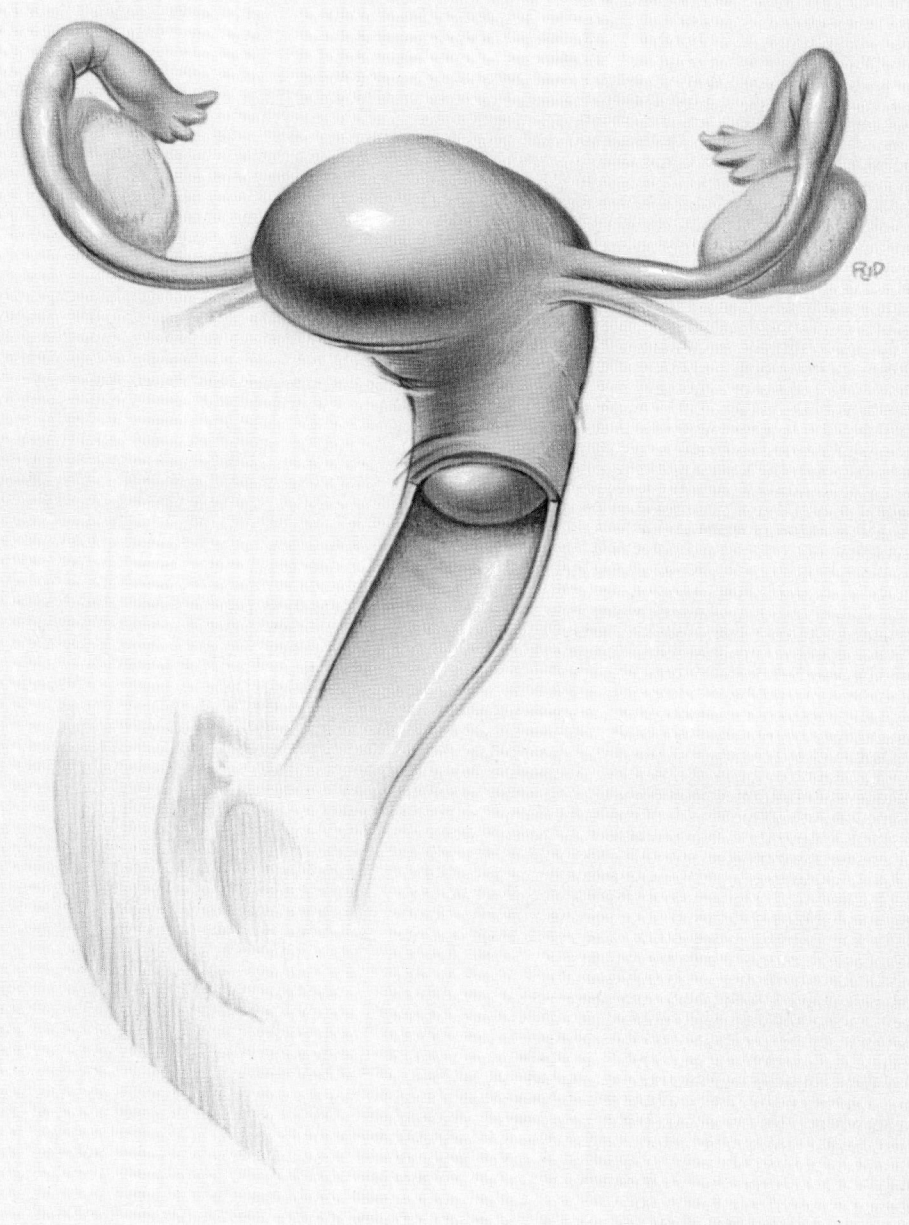

EPILOGUE

Since the beginning of man the biology of conception and reproduction has not changed. The events, from fertilization to birth, reported in this book are constant. They are constant for man, regardless of race, creed, colour, or national origin.

What has changed, however, is man's new-found knowledge of the physiology of human reproduction. For now doctors know about the physiology of pregnancy. They are aware of the mechanics of birth and can help the mother during labour.

With this knowledge man has been able to develop a wide range of methods, devices and chemicals, which afford reasonably predictable contraception and good medical care during pregnancy. Thus, with the contraceptive measures available to them, and with a complete understanding of human reproduction, doctors are able to counsel couples on their contraceptive needs and help them to plan the frequency of pregnancies and the size of their family.

GLOSSARY

ABORTION. The termination of pregnancy during the first twenty-eight weeks of pregnancy.

AFTERBIRTH. The placenta and foetal membranes expelled after delivery of the baby.

AMNION. A thin, transparent membranous sac which forms the inner wall of the foetal membranes and which holds the foetus suspended in fluid.

"BAG OF WATERS." The foetal membranes. See amnion.

BASAL BODY TEMPERATURE GRAPH (BBT). A graph which records the daily body temperature of the female. A rise in temperature indicates that ovulation has occurred.

BIRTH CONTROL. Any method used to control pregnancy.

BLASTOCYST. A stage in the development of an embryo. In this stage, the embryo consists of a layer of cells surrounding a fluid-filled cavity.

BULBOURETHRAL GLANDS. Two small glands located on each side of the prostate gland. They secrete a fluid forming part of the seminal fluid. Also called Cowper's glands.

CALENDAR RHYTHM METHOD. A contraceptive technique which allows intercourse during that period of the menstrual cycle when ovulation is not likely to occur. This technique uses calendar days for calculation.

CERVICAL CAP. A device placed over the cervix to prevent conception.

CERVICAL MUCUS. An adhering, slippery secretion of the cervix.

CERVIX. The narrow, outer end of the uterus resembling a neck.

CAESAREAN SECTION. The delivery of a baby by way of a surgical incision of the walls of the abdomen and uterus.

"CHANGE OF LIFE." See menopause.

CHORION. The outer layer of the foetal membranes which serves as a protective and nutritive covering for the fertilized egg, the developing embryo, and the foetus.

CHORIONIC GONADOTROPHIN. A hormone produced by the placenta which stimulates the *corpus luteum* of the ovary to produce progesterone.

CHORIONIC SAC OR VESICLE. See chorion.

CHROMOSOMES. Small rod-shaped bodies present in the cell nucleus which contain hereditary factors.

CIRCUMCISION. The removal of the foreskin of the glans penis.

CLITORIS. A small organ located at the anterior of the external female genital organs.

COITUS. Sexual intercourse between persons of the opposite sex.

COITUS INTERRUPTUS. Withdrawal of the penis from the vagina before the ejaculation of semen.

COMBINATION PILLS. Birth control pills that contain two synthetic hormones (oestrogen and progesterone) in the same tablet. When these pills are used, the endometrium and cervical mucus approximate the condition that they would have during the postovulatory phase of the menstrual cycle.

CONCEPTION. The union of the sperm and the egg, which results in pregnancy.

CONDOM. A covering worn over the penis during intercourse to prevent pregnancy.

CONTRACEPTION. The voluntary prevention of conception or pregnancy.

CORPUS LUTEUM. A yellow mass in the ovary formed by an ovarian follicle that has matured and discharged its egg.

COWPER'S GLAND. See bulbourethral glands.

DELIVERY. Passage of the baby through the birth canal.

DIAPHRAGM. A rubber or plastic cup which fits over the cervix and is used for contraceptive purposes.

DOUCHE. A hygienic procedure consisting of passing a stream of solution into the vagina.

DYSMENORRHOEA. Painful or uncomfortable menstruation.

ECTOPIC PREGNANCY. The abnormal implantation of the fertilized egg outside the uterine cavity.

EGG. The female cell.

EJACULATION. The expulsion of semen from the male urethra.

EJACULATORY DUCTS. The terminal portion of the duct through which sperm are conveyed to the urethra.

EMBRYO. The stage of early development when the organs of the body are being formed.

EMISSION. A discharge of semen into the urethra.

ENDOMETRIUM. The tissue layer lining the inner surface of the uterus.

EPIDIDYMIS. An elongated structure at the back of the testis, in which the sperm are stored.

EPISIOTOMY. Surgical incision of the vulva during childbirth.

ERECTILE TISSUE. Vascular tissue which, when filled with blood, becomes erect or rigid, such as the clitoris or penis.

ERECTION. The state of swelling, hardness, and stiffness observed in the penis, and to a lesser extent in the clitoris, due to sexual excitement.

FALLOPIAN TUBE. The tube or duct which extends laterally from the uterus, terminating near the ovary. It conducts the egg from the ovary to the uterus and the sperm from the uterus toward the ovary.

FERTILE PERIOD. That period during which the female is most apt to become pregnant.

FERTILIZATION. The union of sperm and egg.

FOETUS. An unborn child between the fourteenth week of pregnancy and birth.

FORESKIN. Loose skin at and covering the end of the penis.

FRATERNAL TWINS. Twins which are the product of two separate fertilized eggs and are not identical.

GENITAL. Relating to the organs of reproduction in both the male and female.

GLANS PENIS. Bulbous end or head of the penis in which the urethral orifice is located.

GONAD. The female sex glands or ovaries and the male sex glands or testes.

GONADOTROPHIC HORMONE. See gonadotrophin.

GONADOTROPHIN. A hormone which stimulates the sex glands.

HORMONE. A chemical substance produced by a gland and carried by the blood stream to another area of the body where it exerts its effect.

HYMEN. A fold of tissue partly closing the vaginal opening.

HYPOTHALAMUS. The area of the brain adjacent to the pituitary gland.

IDENTICAL TWINS. Twins which are the product of one single fertilized egg and are of the same genetic makeup and sex.

IMPLANTATION. Attaching and embedding of the blastocyst into the uterine wall.

INTROITUS. The opening of the vagina.

INTRAUTERINE DEVICE (IUD/IUCD). A plastic or metal device inserted in the uterus as a means of contraception.

LABIA MAJORA. The two rounded folds on either side of the vulva.

LABIA MINORA. The two thin folds of skin lying on either side of the vulva between the labia majora and the opening of the vagina.

LABOUR. The physiological process by which the foetus is expelled by the uterus at time of birth.

LACTATION. The secretion of milk from the female's breasts.

LACTOGEN. A hormone which stimulates the secretion of milk.

LUTEINIZING HORMONE. A gonadotrophic hormone which is secreted by the pituitary gland and causes the ovarian follicle to rupture and release an egg.

MATURATION. The process of becoming mature.

MENOPAUSE. That period which marks the permanent stoppage of menstrual activity.

MENSTRUAL CYCLE. The reproductive cycle of the human female characterized by a recurrent series of changes in the uterus and sex organs.

MENSTRUATION. The periodic discharge of a bloody fluid from the uterus into the vagina at regular intervals during the menstrual cycle.

MID-TRIMESTER. The middle three months of pregnancy.

MINI PILLS. An oral chemical agent still in the experimental phase which affects the cervical mucus and blocks sperm penetration.

MISCARRIAGE. See abortion.

MITTELSCHMERZ. Pain between menstrual periods, associated with ovulation.

MORULA. An early stage in development during which the fertilized egg consists of a solid mass of cells, resembling a mulberry.

NUTRIENT. Food for the body.

NOCTURNAL EMISSION. Spontaneous ejaculation during sleep.

OESTROGEN. A female sex hormone produced by the ovarian follicle which stimulates the internal female reproductive organs and the development of secondary sex characteristics in the female.

ORAL CONTRACEPTIVE. A method of preventing conception based on oral medication or pills.

ORGASM. A state of excitement which occurs during sexual intercourse.

OVARIAN FOLLICLES. A spherical structure in the ovary consisting of an egg and its surrounding cells.

OVARIES. Two glands in the female producing the eggs and hormones.

OVIDUCTS. Two tubes extending from the uterus which convey the egg from the ovary to the uterus. See fallopian tube.

OVULATION. The ripening of the mature ovarian follicle and the release of the egg.

OXYTOCIN. A pituitary hormone which stimulates contraction of the uterine masculature to expel the foetus.

PELVIS. The bones of lower portion of the trunk of the body.

PENIS. The male organ.

PERINEUM. The area at the base of the body between the thighs.

PITUITARY GLAND. A small structure at the base of the brain which secretes hormones.

PITUITARY HORMONE. A hormone secreted by the pituitary gland.

PLACENTA. The oval spongy structure in the uterus through which the foetus derives its nourishment. Major component of the afterbirth.

POLAR BODY. One of two minute cells given off successively by the egg during its development.

POSTCOITAL PILLS. A means of contraception with pills taken after intercourse.

POSTOVULATION. That period after ovulation.

POSTPARTUM PERIOD. That period after the birth of the baby until the mother's body returns to its prepregnant condition.

PREGNANCY. The condition of being with child.

PREMATURE INFANT. A child born before full development.

PREOVULATION. The period before ovulation.

PREPUCE. The fold of skin that covers the end of the penis or the glans penis.

PROGESTERONE. A female sex hormone produced by the *corpus luteum* of the ovary.

PROSTATE GLAND. A gland located around the male urethra, which secretes a thin fluid that forms part of the semen.

PUBERTY. That time of life when the male or female becomes capable of reproduction.

REPRODUCTION. The process which gives rise to offspring.

REPRODUCTIVE SYSTEM. The organs in the male and female that are utilized in the role of creating new life.

RHYTHM METHOD. A contraceptive method which allows sexual intercourse only during the time of the menstrual cycle that the female is least susceptible to fertilization.

"SAFE" PERIOD. The time during the menstrual cycle when the female is least fertile.

SCROTUM. The skin pouches beneath the penis containing the testicles.

SEMEN. A thick milky secretion composed of sperm and seminal fluid.

SEMINAL POOL. The deposit of semen in the vagina.

SEMINAL VESICLES. Saclike structures in the male which secrete a thick fluid that forms part of the semen.

SEQUENTIAL PILLS. An oral contraceptive medication packaged so that the woman receives oestrogen for three-quarters of her cycle and oestrogen and progesterone for the remainder. This type of pill produces activity that parallels the normal cycle.

SEXUAL INTERCOURSE. Union of the male and female in which the penis is introduced into the vagina.

SPERM. The male fertilizing cell.

SPERMICIDAL PREPARATION. A chemical substance that will kill or immobilize sperm.

SPERMATOCYTE. A cell which will give rise to spermatozoa.

SPERMATOGENESIS. The formation of mature functioning sperm in the testes.

SPERMATOZOA. See sperm.

STERILIZATION. A procedure whereby a male or female is rendered incapable of reproduction. See tubal ligation and vasectomy.

SYNTHETIC HORMONE. A chemical manufactured artificially to approximate that produced naturally.

TEMPERATURE RHYTHM METHOD. A contraceptive technique which allows intercourse during that period of the menstrual cycle when ovulation is not likely to occur. This technique uses basal body temperature for calculation. See basal body temperature graph.

TERM FOETUS. An unborn child between thirty-eight and forty weeks of pregnancy.

TESTES. The male reproductive glands located in the scrotal sacs and which produce sperm and the hormone testosterone.

TESTOSTERONE. The principal male hormone which accelerates the development of secondary sexual characteristics. It is essential for normal sexual behaviour and is responsible for the deepening of the male voice, muscle development, and growth of beard and pubic hair.

TRIMESTER. A period of three months. A full-term pregnancy of nine months is divided into three trimesters.

TROPHOBLAST. The outer layers of a developing fertilized egg that will become the placenta and foetal membranes.

TUBAL LIGATION. Female sterilization in which the fallopian tubes are cut and the middle section usually removed.

UMBILICAL CORD. The cord connecting the foetus with the placenta.

URETHRA. The canal that discharges urine.

URINARY BLADDER. The receptacle that holds the urine before it is excreted.

URINARY MEATUS. The external opening of the urethra.

UTERINE LINING. The tissue covering the inside of the womb.

UTERUS. The muscular organ of the female which houses the foetus. Also known as the womb.

VAGINA. The canal extending from the vulva to the cervix.

VAS DEFERENS. The duct from the testes that carries the sperm.

VAS LIGATION. See vasectomy.

VASECTOMY. Male sterilization which involves surgical cutting of the *vas deferens* to cause permanent sterility.

VERNIX CASEOSA. A fatty secretion covering the foetus, most prevalent in the creases and folds of the skin.

VULVA. External parts of the female genital organs.

WITHDRAWAL. The removal of the penis from the vagina before ejaculation in order to prevent introduction of semen into the female.

WOMB. See uterus.

ZONA PELLUCIDA. A layer of material surrounding the egg through which the sperm must pass before fertilization can take place.